Urban Regeneration

Urban Regeneration
A Handbook

Edited by
PETER ROBERTS AND HUGH SYKES

SAGE Publications

London · Thousand Oaks · New Delhi

First published 2000

SAGE Publications Ltd
6 Bonhill Street
London EC2A 4PU

SAGE Publications Inc.
2455 Teller Road
Thousand Oaks, California 91320

SAGE Publications India Pvt Ltd
32, M-Block Market
Greater Kailash-I
New Delhi 110 048

British Library Cataloguing in Publication Data
A catalogue record for this book is
available from the British Library

ISBN 0 7619 6716 8
ISBN 0 7619 6717 6 (pbk)

**Library of Congress catalog card number available from the
publisher**

Typeset by Dorwyn Ltd, Rowlands Castle, Hants
Printed and bound in Great Britain by the Cromwell Press, Wilts

Contents

Figures, Tables and Boxes

Figures

Tables

Boxes

Contributors

The Editors

Peter Roberts is Professor of European Strategic Planning at the University of Dundee. He is Chair of the British Urban Regeneration Association Best Practice Committee, Chair of the Town and Country Planning Association and an Adviser to the Local Government Association. He has published on a wide range of topics, including urban and regional development and regeneration, European regional policy and planning, spatial development, environmental management, and the politics of devolution.

Sir Hugh Sykes DL was Chairman of the Sheffield Development Corporation from its formation in 1988 until it ceased operations in 1997. He currently serves on the boards of a number of public companies, including Yorkshire Bank plc. He is a past Master of the Company of Cutlers in Hallamshire and is Treasurer of the University of Sheffield.

The Contributors

Amanda Beresford is a solicitor and a qualified planner. She is head of Environment Law at Addleshaw Booth & Co.

Andrew Carter is Regeneration Strategies Adviser at Greater London Enterprise. He is a member of BURA's Best Practice Committee and Director of the Urban Forum.

Paul Drewe is Professor of Spatial Planning at Delft University of Technology and Visiting Professor of Urban Management at the University of Ghent. He also works as a consultant both in the Netherlands and abroad.

Clive Dutton OBE is Head of Regeneration at Sandwell Metropolitan Borough Council. He currently serves as an adviser to the British government's Urban Task Force.

Martin Eagland has direct experience of implementing urban regeneration schemes, most notably as Chief Executive of the former Leeds Development Corporation. He is now the Managing Director of Eagland Planning Associates.

Bill Edgar is Senior Lecturer at the University of Dundee and a Director of the Joint Centre for Scottish Housing Research. He is also the Research

Co-ordinator of the European Observatory on Homelessness for FEANTSA.

Richard Fleetwood is a partner in Addleshaw Booth & Co., specialising in corporate matters, including public–private partnerships and joint ventures.

Mark Gaffney is a solicitor at Addleshaw Booth & Co., specialising in commercial and residential development, mainly for developers.

Trevor Hart is a Lecturer in town planning and housing at Leeds Metropolitan University. He has a background in local government and consultancy where he was principally concerned with economic development.

Brian Jacobs is Reader in Public Policy in the School of Health at Staffordshire University. He works as a consultant to both public and private organisations and he has published extensively on international urban and regional policy issues.

Paul Jeffrey is a Director of ECOTEC Research and Consulting. He has extensive experience of urban planning, development and regeneration.

Ian Johnston is Principal and Vice-Chancellor at Glasgow Caledonian University and was formerly Director-General of Training, Education and Enterprise at the Department of Employment.

Dalia Lichfield is a partner in Dalia and Nathaniel Lichfield Associates, a development planning consultancy specialising in integrated planning, regeneration and impact assessment. She is a member of BURA, the UK Evaluation Society and the Urban Villages Project Committee.

Barry Moore is Assistant Director of Research at the Department of Land Economy, University of Cambridge. He is also a fellow at Downing College, Cambridge, Research Associate at the ESRC Centre for Business Research and Senior Adviser at Public and Corporate Economic Consultants (PACEC).

David Noon is Associate Dean in the Coventry Business School and Deputy Director of the Centre for Local Economic Development, both at Coventry University. He is also a member of the BURA Best Practice Committee.

John Pounder is a research manager at ECOTEC Research and Consultancy. He has considerable experience of urban regeneration and of economic development.

John Shutt is Eversheds Professor of Regional Business Development at Leeds Metropolitan University. He directs the European Regional Business and Economic Development Unit in the Leeds Business School.

James Smith-Canham is a consultant with KPMG, Birmingham. Formerly he was a lecturer at Coventry Business School, Coventry University.

Rod Spires is the Director of Public and Corporate Economic Consultants. He is also an adviser to English Partnerships and was recently Research Fellow at the Department of Land Economy, University of Cambridge.

John Taylor was Chief Executive of BURA from 1994 to 1998. He chaired the Assessment Panel for the Secretary of State's Awards for Partnership in Regeneration. He is currently Director of the Buildings Crafts College, based in central London.

Foreword

There has been a sea change in the UK in the last two years. Increasingly, regeneration has become a centrepiece of community and economic regeneration not just in the UK, but across Europe.

Since the birth of the British Urban Regeneration Association (BURA) in 1990, the understanding of, and issues surrounding, regeneration have grown in importance and complexity.

It seemed for a while, however, as if BURA was a voice in the wilderness. It attempted to ensure that everyone in what we now call regeneration was able to grasp that merely dealing with the built environment, and not the people involved, would simply be rearranging the deckchairs on the *Titanic*.

There was, at no stage, any guide or help along the way.

This book is BURA's first attempt to fill that gap.

It has been written by renowned experts in their respective fields. Each chapter is a treatise in and of itself, allowing the reader to explore the issues and read further on the subject. The book places into context the route from which current thinking on regeneration came; it identifies key developments along the way.

The book looks elsewhere as well; the UK, the USA and Europe all feature. Theory and practice mix across the book, allowing the reader to develop both knowledge and expertise. It develops ideas and ability, both of which will help the reader searching for help in dealing with the thinking behind the 1999's imperatives of the Urban Task Force, the proposed Urban White Paper and more besides.

As Dick Caborn, Minister for Planning, Regeneration for the Regions, makes clear in the Preface, this book is 'a valuable aid to all those involved' in regeneration.

The unique blend of the private, public and not-for-profit sectors places BURA at the heart of the new agenda for regeneration. It is from this rich mix that we were able to draw together such a wealth of expertise.

The acknowledgements are extensive. The British Urban Regeneration Association owes a debt of gratitude to each of the contributors who have worked so hard to hone their work. The publisher has been both patient and helpful. It was gracious of the Minister to provide such a glowing Preface. Finally, the book would not have come to fruition without the hard work and commitment of the two Editors. Professor Peter Roberts, collaborating with Sir Hugh Sykes, has made this book the success that it is. Peter has managed to do this whilst juggling the Chair of BURA's Best

Practice Committee and the Chairmanship of the Town and Country Planning Association, as well as his real job. Sir Hugh's continued support has itself made this book possible.

I commend this, the first edition of BURA's book, *Urban Regeneration*, to you. Its 'date' is set at mid-1998. The British Urban Regeneration Association would welcome feedback; the developments of the last year already beg that we produce a second edition. Watch this space.

David Fitzpatrick
Chief Executive, BURA

Preface

Regeneration involves the public, private and community and voluntary sectors working together towards a clear single aim – to improve the quality of life for all. We need to ensure that individuals and organisations are learning from the successes – and the failures – of others. This demands a solid core of best practice.

To achieve real and lasting change in our towns and cities, we must make best use of all the resources we have – human and financial. We have put in place a new framework for delivering regeneration, recasting the Single Regeneration Budget and setting up the New Deal for Communities with funding of £800 million over the next three years. Currently 17 pathfinder partnerships are working on new ways of tackling the entrenched problems of some of the country's most disadvantaged neighbourhoods.

These new partnerships will be looking at how they can build on current practice and how they can put the local community at the heart of partnership. We expect them to be innovative, pushing forward the boundaries of what is achievable.

BURA's work in identifying and spreading best practice is making a valuable contribution to the regeneration process. This book offers a comprehensive account of regeneration activity here and abroad, and will be a valuable aid to all those involved in its practice.

Richard Caborn MP
Minister of State for the
Regions, Regeneration and Planning

List of Abbreviations

AMID	Almonaster-Michoud Industrial District
BBO	broad-based organising
BCDC	Black Country Development Corporation
BCNR	Black Country New Road
BICA	Bede Island Community Association
BiC	Business in the Community
BUD	British Urban Development
BURA	British Urban Regeneration Association
CAT	City Action Team
CBD	Central Business District
CBI	Confederation of British Industry
CCC	Center for Community Change
CDC	Community Development Corporation
CDP	Community Development Project
CEB	Community Enterprise Board
CHDC	Citizens Housing Development Corporation
CIE	community impact evaluation
COSLA	Convention of Scottish Local Authorities
DDS	dead-weight displacement and substitution
DETR	Department of the Environment, Transport and the Regions
DFBO	Design Finance Build Operate
DLG	Derelict Land Grant
DoE	Department of the Environment
EP	English Partnerships
ERDF	European Regional Development Fund
ESF	European Social Fund
EU	European Union
EZ	Enterprise Zone
GEAR	Glasgow Eastern Area Renewal
GIS	geographical information system
GNVQ	General National Vocational Qualification
GOR	Government Office for the Regions
HAA	Housing Action Area
HAT	Housing Action Trust
HGV	heavy goods vehicle
HIP	Housing Investment Programme
HPAP	Home Purchase Assistance Program

HRH	Housing Revenue Account
HUD	Housing and Urban Development
IAS	Inner Area Study
ISDN	Integrated Services Digital Network
IT	information technology
JTPA	Job Training Partnership Act
LDDC	London Docklands' Development Corporation
LEC	Local Enterprise Company
LGMB	Local Government Management Board
LISC	Local Initiatives Support Corporation
LPAC	London Planning Advisory Committee
MBC	Metropolitan Borough Council
MHCDO	Marshall Heights Community Development Organisation
NCC	Newark Community Corporation
NCVO	National Council for Voluntary Organisations
NRA	National Rivers Authority
NVQ	National Vocational Qualification
OECD	Organisation for Economic Co-operation and Development
PFI	Private Finance Initiative
PPG	Planning Policy Guidance
RDA	Regional Development Agency
RSA	Regional Selective Assistance
RWJF	Robert Wood Johnson Foundation
SME	small and medium-sized enterprises
SNAP	Shelter Neighbourhood Action Project
SPD	Single Programme Document
SPV	special purpose vehicle
SPZ	Simplified Planning Zone
SRB	Single Regeneration Budget
SRBCF	Single Regeneration Budget Challenge Fund
TEC	Training and Enterprise Council
TEF	Treasury Evaluation Framework
TGLP	Thames Gateway London Partnership
TIZ	Town Improvement Zone
TTWA	travel-to-work area
UDA	Urban Development Area
UDC	Urban Development Corporation
UDG	Urban Development Grant
UPA	Urban Priority Area
UPP	Urban Pilot Project
URG	Urban Regeneration Grant
VFM	value for money

PART 1

THE CONTEXT FOR URBAN REGENERATION

1   Introduction

Peter Roberts and Hugh Sykes

Urban regeneration is a widely experienced but little understood phenomenon. Although most towns and cities have been involved in regeneration schemes, and whilst many development companies, financial institutions and community organisations have participated in one or more such ventures, there is no single prescribed form of urban regeneration practice and no single authoritative source of information.

The aim of this book is to remedy this situation by distilling the evidence of good practice and by combining this evidence with explanations of why urban regeneration is necessary and how it functions. A mixture of explanation, evidence and the experience of implementation provides the practical philosophy which has guided the preparation of this book. The intention is to offer the reader a guide to urban regeneration which is comprehensive, accessible and practical. In particular, the book aims to provide an insight into the reasons for the occurrence and persistence of urban problems, the successive changes that have occurred in the practice of urban regeneration and the lessons of good practice.

One of the major difficulties encountered in preparing this book was the virtual absence of quality literature that encompassed the whole of the organisation and functioning of the urban regeneration process. Despite the presence of a wide array of fragmented information on 'fashionable' topics such as partnership, tackling social exclusion, promoting flagship projects, urban greening and how to guarantee success in obtaining 'millennium money', little written material is available that combines coverage of all of the fundamental topics, such as the physical, economic, social and environmental dimensions of regeneration, with the implementation, management and evaluation of the urban regeneration process. This book attempts to remedy this situation. It offers guidance, based on both the theory and the practice of urban change and regeneration, that should prove to be of assistance to those who are engaged in a variety of policy areas and in the active management of urban transition.

The Structure of this Book

The material contained within this book is organised in a way that allows the reader to dip into those sections that are of particular interest.

Although each part and chapter is self-contained to the extent that it deals with a particular theme or subject, the material presented is organised in a manner which allows the reader to gain a rapid overview of the enormous span of urban regeneration issues and activities. Even though the scope and content of this book is wide-ranging in an attempt to provide a comprehensive treatment of the full span of urban regeneration, it would be wrong to suggest that it is a complete treatment of a subject that is so extensive in terms of its practices and applications that it is difficult to define it with any degree of precision. Because urban regeneration is by its very nature a dynamic rather than static phenomenon, it is almost impossible to capture all of the features of current practice or to predict the future with any degree of certainty.

In order to assist the reader, and to set the context for the remainder of the book, an introduction to the origins, challenges and purposes of urban regeneration is presented in Chapter 2. The material in Chapter 2 is cross-referenced to later chapters in order to guide the reader to the more detailed discussions which they contain. Chapter 3 introduces the reader to basic notions such as partnership, strategy and the lessons that may be gained from the study of best practice. These are recurrent features that can be seen in many aspects of urban regeneration practice, and the analysis attempts to identify common elements which help bind together the diverse subjects that are contained in the following chapters. Chapter 2 chiefly provides an introduction to the individual topics that are considered in Part 2, whilst Chapter 3 introduces the management issues contained in Part 3.

Most of the chapters in this book have been prepared by a team of authors who between them represent the required blend of practice experience and academic explanation considered necessary to tackle the complexities that are inherent in any individual aspect of urban regeneration. In Part 2, the contributions of the various teams of authors have been organised in such a way as to provide the reader with an introduction to each of the basic 'building block' themes and topics that are fundamental to an understanding of the urban regeneration process. These chapters deal with:

- economic and financial issues;
- physical and environmental aspects of regeneration;
- social and community issues;
- employment, education and training;
- housing issues.

Cutting across all attempts to stimulate urban regeneration are a number of other important issues. These issues govern the ways in which urban regeneration proceeds and how it is organised. Three 'cross-cutting' issues that are of particular importance in all urban regeneration schemes are examined in Part 3 of this book:

- the legal and institutional basis for regeneration by land development;
- the monitoring and evaluation of regeneration programmes;

- questions of organisation and management.

In order to provide lessons from best practice and offer examples of how to construct and implement strategies for regeneration, the chapters in this book contain a variety of case studies. Other valuable experience can be gleaned from the experience of urban regeneration in countries outwith the UK. The first two chapters of Part 4 offer an insight into some of the major features and important characteristics of efforts to promote urban regeneration in the towns and cities of the mainland of Europe and in North America. A final chapter is also contained in Part 4; the purpose of this chapter is to distil the major lessons from the past and present experience of urban regeneration, to identify the sources of strength and weakness which are evident from such experience, and to propose an agenda for the future. This concluding chapter draws upon the analysis contained in the earlier parts of the book in order to clarify the future role of, and prospects for, urban regeneration as it enters a new century. In addition, the final chapter also considers the extension of urban regeneration to the metropolitan and regional levels.

At the end of each chapter, except in the case of the present chapter and Chapter 14, a set of summary points is provided. These points either indicate some of the key issues and actions arising from the discussion, or provide some key contacts and sources of further information.

Most books of this nature are selective. Other authors and editors would select different themes and cross-cutting issues for inclusion in a volume on the subject of full span regeneration. This book inevitably reflects the skills, experience and preferences of the editorial team and the individual authors: this combination of factors provides the rationale for the selection of material presented here.

Next Steps

Whilst it is apparent that a book of this nature can only ever expect to provide an introduction to urban regeneration theory and practice at a given moment in time – in this case the middle of 1998 – we hope that the first edition will offer guidance and advice to those who are embarking upon the task of regenerating urban areas. The value of such a book is that it can provide immediate help and support, and also stimulate the exchange of experience. It is likely that your experience of urban regeneration will confirm some of the messages contained within this book and it is certain that the material contained in the book will also suggest new ways of approaching difficult and complex problems. We welcome your response to the contents and style of this book and, in addition, we seek your experience – both successes and failures – in order to help us in preparing future editions.

As editors we have gained considerable knowledge and understanding about the subjects addressed in this book during the course of its preparation. We have come to realise how daunting the task of urban regeneration must appear to many who participate in it, and we have discovered that what may seem to be self–evident to one participant in the regeneration process may never have occurred to another. Most import- antly, we have come to appreciate the need to view urban regeneration as a continuous process. No sooner has one problem been solved, than another emerges.

This suggests that it is essential to view the process of urban regeneration as a long-term cycle of activity, there are no 'quick fixes' or permanent solutions here. Each generation faces its own particular set of problems, has its own priorities and works in ways which reflect these priorities. However, whilst each successive generation will face its own particular challenges, the value of learning from previous experience cannot be de- nied. We hope that this book will help to document our state of knowledge in the late 1990s and that it will provide a basis for good practice in urban (and regional) regeneration during the coming years.

Too much time and energy has been lost in the past through 'reinventing the wheel' or through the needless destruction of expert teams that are, in the British way, discarded as one policy initiative and structure succeeds another. This book will have served its purpose if this negative approach is avoided and the accumulated experience of urban regeneration is captured for all to use. The importance of this task cannot be emphasised enough; most policy cycles last a relatively short length of time and the wheel of urban and regional regeneration policy has turned full circle twice during the past 50 years.

In addition, we realise that the institutional and spatial frameworks for regeneration will vary both over time and between places, reflecting both the policy preferences and priorities of government and the perception of the span of the field of action within which regeneration problems can best be addressed. Thus, for example, whilst much urban regeneration effort in the mid-1980s was directed at individual problem sites and small areas, the emphasis in the late 1990s has shifted to the regional level, to communities and to soft infrastructure, to people more so than places. What this tells the keen observer or practitioner, is that regeneration problems and oppor- tunities should best be considered within a spatial continuum. The spectrum of regeneration activity varies from the individual site to the nation-state; there is no single or fixed field of action that represents the ideal spatial level for the practice of regeneration over time.

Looking Forward

It is evident at a late stage in the publication of this book that some of the material is already out of date or is close to becoming time expired. However, there is much contained in the book that is enduring and represents good or best practice irrespective of the specific detail of an individual policy initiative.

In order to assist the reader, the main areas of policy development that are of particular relevance to urban regeneration include the following:

- in relation to the future development of regeneration strategy, the establishment of the Regional Development Agencies (with their strategic role at the regional and subregional levels) and the work undertaken by the Urban Task Force (this was established in 1998 by the Deputy Prime Minister and was charged with identifying the causes of urban decline and recommending practical solutions to bring people back into the cities and urban areas);
- in relation to economic and financial aspects of regeneration, the work of the Regional Development Agencies, the revised funding arrangements and the provision of resources both under traditional programmes, such as Assisted Area policy, and new initiatives including the New Deal;
- in relation to physical and environmental regeneration, a greater emphasis on urban design and quality, an enhanced commitment to sustainable development and the provision of additional resources for the reuse of brownfield land;
- in relation to social and community issues and to employment and training aspects of regeneration, the increased level of resources made available for social housing, regeneration and, especially, the New Deal for Communities;
- various associated advances in policy and practice can be identified, including an increased level of emphasis on local democratic accountability, the provision of regional chambers, devolution to Scotland, Wales and Northern Ireland (and a series of consequential developments that are intended to allow for new policy initiatives and the fine-tuning of existing policies), and the redefinition of a number of key policy objectives in relation to transport, housing, employment, education, planning, environmental management, health, community development and other aspects of regeneration at both urban and regional scales.

Whilst these issues are not dealt with in any detail herein, the material contained in the following pages provides the basic tools that are required in order to design and implement regeneration strategies. Although the details of policy may vary over time, sufficient supplementary literature exists to allow the reader to project forward from the position stated in this book. We will, of course, seek to incorporate the detail of new aspects of policy in a second edition.

Acknowledgements

We wish to thank all those who have helped in the preparation of this book. The bulk of the task of drawing together our knowledge and understanding of urban regeneration has been undertaken by the authors of the various chapters, and to these colleagues we owe a considerable debt of thanks. All of the authors have been assisted in their task by the willing participation of urban regeneration practitioners and analysts representing the authorities, organisations and communities that are both the agents of urban regeneration and the beneficiaries of change. To all of these participants we wish to express our thanks. We are similarly grateful to the members of our Advisory Panel, to the team at the British Urban Regeneration Association (BURA) and to the staff of Sage Publications; these groups have helped to steer this project through the vast and often uncharted waters of urban regeneration theory and practice and, in the case of the latter, have also guided the editors through the mysteries of publishing. Finally we owe a particular debt of thanks to Richard Aspinall, who has provided secretarial support to the editors and who has performed sterling work in assembling the various manuscripts.

2 The Evolution, Definition and Purpose of Urban Regeneration

Peter Roberts

Introduction

Urban areas are complex and dynamic systems. They reflect the many processes that drive physical, social, environmental and economic transition and they themselves are prime generators of many such changes. No town or city is immune from either the external forces that dictate the need to adapt, or the internal pressures that are present within urban areas and which can precipitate growth or decline.

Urban regeneration is an outcome of the interplay between these many sources of influence and, more importantly, it is also a response to the opportunities and challenges which are presented by urban degeneration in a particular place at a specific moment in time. This should not be taken to suggest that all urban problems are unique to a particular town or city, or that solutions advocated and attempted in the past have little relevance to the circumstances of the current day, but it is the case that each urban challenge is likely to require the construction and implementation of a specific response.

Despite having argued that an individual example of urban regeneration is likely to be particular to a specific place, a number of general principles and models of good practice can be identified. Such lessons from current and previous experience can be applied in order to assist in the development and implementation of approaches to the task of regeneration.

This chapter:

- provides a brief history of the origins of urban problems and policy responses;
- defines urban regeneration and identifies the principles which guide its operation;
- provides an introduction to the theory of urban regeneration;
- identifies the purposes of current urban regeneration;
- outlines the development of urban policy.

The Evolution of Urban Areas and some Key Themes

The purpose of this section is to trace the origins of attempts to identify and resolve urban problems, and to isolate the major features and characteristics of the solutions that have been developed and applied. Whilst it is impossible in the space of a few pages to provide anything but the most superficial commentary on some of the major events in the history of urban areas, the most important contribution of this section is to identify the factors that have influenced the emergence of the modern-day practice of urban regeneration.

Previous eras of urban policy have seen the introduction of many novel and well-intentioned schemes aimed both at the resolution of particular problems within existing urban areas and at the establishment of new settlements within, adjacent to or remote from existing towns and cities. As will be seen in the following paragraphs, whilst some of these policy innovations have been based upon advances in technology, others have resulted from new economic opportunities or from the adoption of attitudes to questions of social justice which recognise the likely consequences of allowing urban problems to continue unresolved. Whilst changes in technical capability, economic opportunity and social awareness have been important factors in determining the pace and scale of urban progress, a number of other issues have exerted a significant influence over the form and functioning of cities. The following paragraphs briefly trace this history and identify five major themes that have dominated previous eras of urban change and policy. These themes are:

- the relationship between the physical conditions evident in urban areas and the nature of the social and political response;
- the need to attend to matters of housing and health in urban areas;
- the desirability of linking social improvement with economic progress;
- the containment of urban growth;
- the changing role and nature of urban policy.

Physical Conditions and Social Response

Urban areas have always performed a wide range of functions. Shelter, security, social interaction, and the sale and purchase of goods and services are among the traditional roles of a town or city. The relative importance of each of these functions has changed over time, and such changes have created new demands for land, floor-space, infrastructure and the provision of a range of accompanying facilities. Not surprisingly, some traditional urban areas, either in their entirety or in particular districts of a town, may discover that a previous function or sectoral specialisation is no longer required and that the facilities associated with this function are now redundant. In addition to the role of urban areas as a location for the human functions of living, working and recreation, the physical structures of towns and cities also represent a massive source of wealth. As Fainstein has observed, the distinction between the use of the built environment for human activity and its market role can be 'summarised as the difference between use and exchange values' (Fainstein, 1994, p. 1). This difference, which is reflected in the evolution of the tension between urban areas as places for human activity and as assets, lies at the heart of a number of urban problems and also helps to define the limits within which solutions can be constructed and applied.

Towns and cities change over time, and this process of change is both inevitable and can be viewed as beneficial. It is inevitable because the operation of the political, economic and social systems constantly generate new demands and present fresh opportunities for economic progress and civic improvement. It is beneficial because, although many may deny it, the very existence of these substantial forces of change creates opportunities to adjust and improve the condition of urban areas. As Mumford argued, 'in the city, remote forces and influences intermingle with the local: their conflicts are no less significant than their harmonies' (Mumford, 1940, p. 4). It is the desire to respond positively to such influences that has caused politicians, developers, landowners, planners and citizens alike to search for an answer to the question of how best to improve and maintain the condition of towns and cities.

The responses made to this challenge have varied over time, mirroring the socio-political and economic values and structures of urban society. In previous centuries, new towns and cities were imposed upon societies and settlements were altered by feudal lords and monarchs with no reference to their pre-existing inhabitants – the 'bastide' towns of Gwynedd, to this day, demonstrate their military and colonial origins (Smailes, 1953). However, and reflecting chiefly the history of the past two centuries, most British towns and cities represent an attempt to create or reorder urban areas in a manner that best serves the requirements of a continually evolving industrial society.

Expanding the boundaries of urban areas, together with an associated increase in the diversity of land uses present within pre-existing built-up areas, has been the typical and dominant response to the need to provide additional space for houses, factories, offices and shops. Although there are many examples of grand schemes of civic renewal and the establishment of new industrial settlements, the Victorian slum 'city of dreadful night' (Hall, 1988, p. 14) was the product of a society that paid insufficient attention to the living conditions of the majority of urban residents. For reasons of public health and a genuine desire to improve urban living conditions, the slums of the nineteenth century were eventually acknowledged as an unacceptable end-product of a process whereby industrialisation had dictated the pace and quality of urbanisation. The belated recognition during the last decades of the nineteenth century of the consequences of unregulated urban growth reflects one of the messages that has been carried forward to the present-day practice of urban regeneration: this is the relationship between urban physical conditions and social response. In Joseph Chamberlain's Birmingham of the 1870s, urban improvement was promoted through a 'civic gospel' (Browne, 1974, p. 7) aimed at eradicating living conditions, which, in Chamberlain's view, had created a situation whereby 'it is no more the fault of these people that they are vicious and intemperate than it is their fault that they are stunted, deformed, debilitated and diseased' (Browne, 1974, p. 30).

Chapter 5 deals with issues related to the physical and environmental condition of towns and cities.

Housing and Health

Following the recognition and acceptance of the link between poor physical conditions and social deprivation, a series of policy interventions emerged in an attempt to improve the living conditions of urban residents. The eradication of disease, the provision of adequate housing, the supply of pure water and the creation of open space were early priorities and these areas of activity have proved to be enduring necessities.

Whilst this second dominant theme, which is still present in urban regeneration, had its origins in the response to the slum conditions of the Victorian era, there is a constant need for physical intervention in order to replace outdated or unsatisfactory dwellings and premises. During the Victorian era *in situ* renewal was common, although in many cases at densities far too high to ensure the permanent improvement in living conditions that was originally anticipated, and this was matched, chiefly due to improvements in transport technology, by rapid suburban growth. In addition, and serving as a reminder to the present day of both the possibility and desirability of creating urban conditions in which social, economic and physical improvements can go hand in hand, there was a growing acceptance of the lessons and benefits to be gained from the enlightened experiments in 'model village' living established at Port Sunlight, Bournville, New Lanark and elsewhere.

Chapter 8 discusses the housing dimension of urban regeneration in greater detail.

Social Welfare and Economic Progress

Whilst it was not always the case that physical renewal alone could provide an answer to the many problems which beset the Victorian city, the public health objective of reducing overcrowding and disease did bring about a gradual improvement in the condition of urban areas. Moving beyond this limited objective, and seeking in addition to create an environment in which a third element of urban regeneration – the enhancement of economic prosperity – could be more closely allied to enhanced social welfare and improved physical conditions, Ebenezer Howard and the Garden City Movement experimented in the creation of communities which combined 'all the advantages of the most energetic and active town life, with all the beauty and delight of the country' (Howard, 1902, p. 15). Although only a limited number of garden cities were constructed according to Howard's original conception – Letchworth (1903) and Welwyn Garden City (1920) – the influence of the Garden City Movement was considerable and what survived from the experiment in the form of the post-1945 new towns was 'the essence of the Howard vision' (Hall, 1988, p. 97).

Suburban growth, and especially the rapid growth which followed the building of suburban railways and the later introduction of the bus and car, was a distinctive feature of the late Victorian era and the first part of the twentieth century, and, as noted above, whilst this escape to the suburbs provided a relief valve for the more affluent and mobile, it did little to relieve the problems of the inner parts of towns and cities. In the period after 1870 most British urban areas acquired a cheap and efficient public transport system, followed later by the introduction and increasing use of the private motor car. The impact on urban growth of these new transport technologies was rapid and widespread. As Hall (1974) notes, up to the 1860s densities in London were rising and the city was contained – the population doubled between 1801 and 1851, but the area of the city did not increase in proportion. However, following the introduction of new transport technologies the city began to spread, especially in the period after 1918 – in 1914 the population of London was 6.5 million, by 1939 8.5 million, whilst the built-up area had trebled in size.

Chapters 4 and 6 provide further information on questions of economic and social change.

Containing Urban Growth

This introduces the fourth theme from the past that can be seen to have influenced and shaped the current purpose and practice of urban regeneration. This theme has its origins in the perceived need to restrain urban growth and to make the best possible use of the land that is already used for urban functions. Urban containment provided a rationale both for the *in situ* renewal of urban areas and for the balanced expansion of settlements beyond the green belts, which were increasingly imposed around the major towns and cities from the 1930s onwards.

Attempts to contain urban sprawl and to ensure the maximum beneficial use of land already within the urban area have dominated much of urban policy during the past century. This theme is still of considerable importance and provides an immediate stimulus for much urban regeneration.

Peter Roberts

Table 2.1   **The evolution of urban regeneration**

Period Policy type	1950s Recon- struction	1960s Revital- isation	1970s Renewal	1980s Redevelop- ment	1990s Regener- ation
Major strategy and orien-tation	Reconstruction and extension of older areas of towns and cities often based on a 'masterplan'; suburban growth.	Continuation of 1950s theme; suburban and peripheral growth; some early attempts at rehab-ilitation.	Focus on *in-situ* renewal and neigh-bourhood schemes; still development at periphery.	Many major schemes of development and redevelop-ment; flagship projects; out of town projects.	Move towards a more com-prehensive form of policy and practice; more emphasis on integrated treatments.
Key actors and stakeholders	National and local government; private sector developers and contractors.	Move towards a greater balance between public and private sectors.	Growing role of private sector and de-centralisation in local government.	Emphasis on private sector and special agencies; growth of partnerships.	Partnership the dominant approach.
Spatial level of activity	Emphasis on local and site levels.	Regional level of activity emerged.	Regional and local levels initially; later more local emphasis.	In early 1980s focus on site; later emphasis on local level.	Reintroduction of strategic perspective; growth of regional activity.
Economic focus	Public sector investment with some private sector involvement.	Continuing from 1950s with growing influence of private investment.	Resource constraints in public sector and growth of private investment.	Private sector dominant with selective public funds.	Greater balance between public, private and voluntary funding.
Social content	Improvement of housing and living standards.	Social and welfare improvement.	Community-based action and greater empower-ment.	Community self-help with very selective state support.	Emphasis on the role of community.
Physical emphasis	Replacement of inner areas and peripheral development.	Some continuation from 1950s with parallel rehabilitation of existing areas.	More extensive renewal of older urban areas.	Major schemes of replacement and new development; 'flagship schemes'.	More modest than 1980s; heritage and retention.
Environmental approach	Landscaping and some greening.	Selective im-provements.	Environmental improvement with some in-novations.	Growth of concern for wider approach to environment.	Introduction of broader idea of environmen-tal sustain-ability.

Sources: After Stöhr (1989) and Lichfield (1992).

Changing Urban Policy

So the scene is now set for a brief description and assessment of the evolution of urban policy over the past half century, and for the identification of the fifth and final theme from the past that has influenced the current theory and practice of urban regeneration. This final element reflects the changing assignment of responsibility for the improvement and management of towns and cities. From post-Second World War reconstruction to the present-day model of partnership, power and responsibility for the discharge of the tasks of urban regeneration has changed hands in line with the broader conventions of social organisation and the dominant forces of political life. The pattern of evolution of urban policy, together with the characteristics of each era of policy, is summarised in Table 2.1.

In the immediate period after 1945 repairing wartime damage and reconstructing the fabric of towns and cities, many of which had been neglected for years, initially took priority. This process of reconstruction was seen as a task of national importance. The pace was set by central government, with the Ministry of Town and Country Planning even offering detailed guidance to local authorities 'on the principles and standards that should govern the preparation of redevelopment plans for (central) areas' (Ministry of Town and Country Planning, 1947, p 1). With detailed guidance of this kind it is little wonder that so many of the end-products of the post-war schemes of central renewal look depressingly alike.

Other policy prescriptions were launched alongside central redevelopment. Urban constraint through the designation of green belts still permitted substantial peripheral expansion within the urban fence, and further suburbanisation also occurred at the edges of many existing towns and cities. Beyond the green belt were the new and expanded towns, together with rapidly growing free-standing county towns. The emphasis in the 1940s and 1950s was on reconstruction, replacement and the eradication of the physical problems of the past. Government-led, with enthusiastic support from local authorities and the private sector alike, the priorities of slum clearance and reconstruction led to the embrace of 'high-rise housing and industrialised building techniques' (Couch, 1990, p. 29).

By the mid-1960s it was already apparent that many of the immediate post-war solutions had simply transferred the location and altered the manifestation of urban problems. Growing dissatisfaction with slum clearance and the resulting decanting of population to peripheral estates, together with a more participatory and decentralised approach to government, led to a series of adjustments to policy. In the urban policy field this shift in priorities resulted in an increased emphasis on improvement and renewal. This 'discovery' of the inner city, together with the first tentative steps towards the generation of urban policy, led to a major expansion of urban initiatives during the 1970s. Associated with the proliferation of initiatives in this period were a series of attempts to ensure greater co-ordination between the previously separate economic, social and physical strands of policy.

Many of the urban policy initiatives of the 1970s initially continued into the 1980s, although substantial modifications and additions were subsequently introduced (Turok, 1987). Most significantly, during the 1980s there was a move away from the idea that the central state should or could provide all of the resources required in order to support policy interventions. This new policy stance was matched by a greater emphasis on the role of partnership. The more commercial style of urban redevelopment evident in the 1980s reflected yet another set of changes in the nature and structure of political philosophy and control.

Further adjustments to the form and operation of urban policy have occurred in the 1990s, with a gradual move back to a more consensual style of politics and the recognition of a series of new problems and challenges. This change in stance has influenced the form and content of urban policy. One example of the new policy formulation of the 1990s, which is evident both in the general domain of politics and in urban policy, is the acceptance of the need to work in accord with the environmental objectives of sustainable development. Although not yet fully reflected in what we now define as urban regeneration, this is a final illustration of the way in which the inheritance of the past and the challenges of the present help to shape urban regeneration. Although the new challenge of environmentally sustainable development has not yet fully imposed its characteristics on the functioning of urban areas, there is little doubt that it is likely to dominate the theory and practice of urban regeneration and of urban management in the future.

The Basis for Urban Regeneration

These themes from the history of urban problems and opportunities: the relationship between physical conditions and social response; the continued need for the physical replacement of many elements of the urban fabric; the importance of economic success as a foundation for urban prosperity and quality of life; the need to make the best possible use of urban land and to avoid unnecessary sprawl; and the importance of recognising that urban policy mirrors the dominant social conventions and political forces of the day, are themes which will be developed elsewhere in this book. At this point these five themes, plus the new theme of sustainable development, simply need to be acknowledged.

As is demonstrated more fully in the following section of this chapter, there is a high degree of coincidence between the history of the content, structure and operation of urban policy, and the general evolution of political attitudes, social values and economic power. However, although the style and characteristics of successive rounds of urban policy reflect the evolution of political, economic and social values, and although particular urban problems and some aspects of urban policy have come and gone over time, a professional and technical capacity has emerged in response to the challenge of urban regeneration. This capacity has continued to evolve almost irrespective of the particular political fashions of the day. From new settlements to suburbanisation, and from comprehensive redevelopment to *in situ* regeneration, the urban challenge continues to tax the ability and ingenuity of policy-makers, planners, developers and citizens alike.

What is Urban Regeneration?

Having identified and traced the evolution of some of the major issues and factors that have been evident in previous eras of urban change and policy, the preceding section of this chapter isolated five important themes that can be identified from the history of urban regeneration. These themes chiefly represent the origins and outcomes of past problems and policy responses, and although they reflect the enduring and continuous nature of economic, social and physical change, they do not yield, by themselves, the basis for a comprehensive definition of urban regeneration. In order to help to construct a working definition of urban regeneration it is also necessary to identify emerging areas of concern and likely future challenges. As was argued above, the most important of these challenges is that which is represented by the need to ensure that all areas of public and private policy operate in accord with the principles of sustainable development; this represents a new sixth theme.

A Definition of Urban Regeneration

These six themes – five from the past and one representing the dominant policy issue of the present and future – provide the basis for an initial definition of urban regeneration as:

> comprehensive and integrated vision and action which leads to the resolution of urban problems and which seeks to bring about a lasting improvement in the economic, physical, social and environmental condition of an area that has been subject to change.

This definition encompasses the essential features of urban regeneration that have been identified by Lichfield, where she points to the need for 'a better understanding of the processes of decline' and an 'agreement on what one is trying to achieve and how' (Lichfield, 1992, p. 19); by Hausner, who emphasises the inherent weaknesses of approaches to regeneration that are 'short-term, fragmented, ad hoc and project-based without an overall strategic framework for city-wide development' (Hausner, 1993, p. 526); and by Donnison in his call for 'new ways of tackling our problems which focus in a co-ordinated way on problems and on the areas where those problems are concentrated' (Donnison, 1993, p. 18).

Urban regeneration moves beyond the aims, aspirations and achievements of urban renewal, which is seen by Couch as 'a process of essentially physical change' (Couch, 1990, p. 2), urban development (or redevelopment) with its general mission and less well-defined purpose, and urban revitalisation (or rehabilitation) which, whilst suggesting the need for action, fails to specify a precise method of approach. In addition, urban regeneration implies that any approach to tackling the problems encountered in towns and cities should be constructed with a longer-term, more strategic, purpose in mind.

Bringing together the evidence from the history of urban change and policy with the points made in the previous paragraph, it is possible to identify a number of phases or stages in the development of the theory and practice of what we now define as urban regeneration. Building on the work of Stöhr (1989) and Lichfield (1992), Table 2.1 traces some of the major changes that have occurred in the approach to, and content of, urban policy and practice from the 1950s to the present day.

Principles of Urban Regeneration

Building on the definition provided above, a number of principles can be identified that are the hallmark of urban regeneration. Reflecting the challenges of urban change and their outcomes, which were discussed in the previous section of this chapter, urban regeneration should:

- be based upon a detailed analysis of the condition of an urban area;
- be aimed at the simultaneous adaptation of the physical fabric, social structures, economic base and environmental condition of an urban area;
- attempt to achieve this task of simultaneous adaptation through the generation and implementation of a comprehensive and integrated strategy that deals with the resolution of problems in a balanced, ordered and positive manner;
- ensure that a strategy and the resulting programmes of implementation are developed in accord with the aims of sustainable development;
- set clear operational objectives which should, wherever possible, be quantified;
- make the best possible use of natural, economic, human and other resources, including land and existing features of the built environment;
- seek to ensure consensus through the fullest possible participation and co-operation of all stakeholders with a legitimate interest in the regeneration of an urban area; this may be achieved through partnership or other modes of working;
- recognise the importance of measuring the progress of strategy towards the achievement of specified objectives and monitoring the changing

nature and influence of the internal and external forces which act upon urban areas;

• accept the likelihood that initial programmes of implementation will need to be revised in-line with such changes as occur;

• recognise the reality that the various elements of a strategy are likely to make progress at different speeds; this may require the redirection of resources or the provision of additional resources in order to maintain a broad balance between the aims encompassed in a scheme of urban regeneration and to allow for the achievement of all of the strategic objectives.

Figure 2.1 provides an illustration of the interaction between these and many other factors. This diagram also indicates the variety of themes and topics involved in urban regeneration and the multiplicity of interrelated outputs.

These principles put substance behind the definition of urban regeneration which was given earlier. Above and beyond these principles is the need to recognise and accept the uniqueness of place – Robson expresses this as the 'uniqueness of how things happen in a local area' (Robson, 1988, p. ix) – and the requirement for any particular model of urban regeneration to be calibrated to the circumstances within which it operates. This implies, for example, that an individual scheme of urban regeneration should both reflect the wider circumstances and requirements of the city or region in which it is located (Hausner, 1993), and seek to reduce social exclusion and enhance the economic reintegration of disadvantaged urban areas (McGregor and McConnachie, 1995).

Above and beyond these requirements, which support the principles of urban regeneration stated above, is the desirability of ensuring that urban areas make a positive contribution to national economic performance and to the attainment of a range of other social and environmental goals. In the past some observers have argued that disadvantaged urban areas, and in particular the inner cities, act as a drag upon national and regional success and should be abandoned, but the evidence for such a stance is, at best, flimsy. More recent assessments dismiss the view that disadvantaged inner urban areas should be abandoned because they are no longer important to the success and prosperity of the regions and nations in which they are located. This point has been expressed with force on both sides of the Atlantic. Stegman notes that 'the tragedy of the inner city affects everyone' and that the 'overall performance of metropolitan regions is linked to the performance of their central cities, and urban distress moves outwards from the core' (Stegman, 1995, p. 1602). In essence, what Stegman and others are saying is that cities matter, and that the task of ensuring the effective regeneration of an urban area is of fundamental importance to a wide range of actors and stakeholders, including local communities, city and national government, property owners and investors, economic activities of all kinds, and environmental organisations at all levels from the global to the local.

Figure 2.1   **The Urban Regeneration Process**

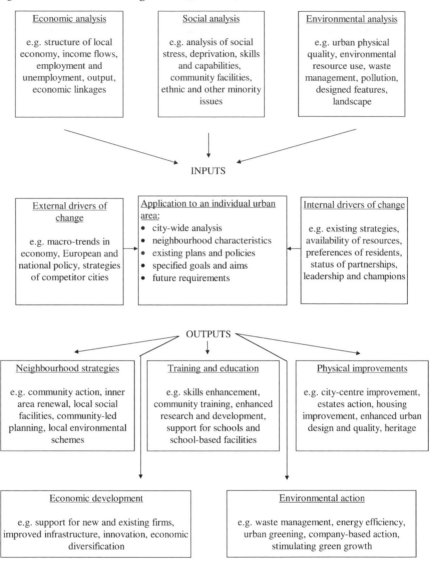

From Theory to Practice

This section of the chapter offers a brief review of some of the major theories which provide a foundation for the practice of urban regeneration. Two immediate problems here are the absence of a single accepted theory that is capable of explaining the entire range of issues related to the occurrence and outcomes of urban change, and the existence of widely differing views as to what constitutes the scope and competence of urban regeneration.

Most explanations of the process of urban change commence their analysis from the consideration of a single factor. They then seek to widen their scope by reference to the outcomes of urban change, rather than the underlying causes. The end result is that most theories of urban change provide only a partial insight into what is a complex process.

An additional issue that has to be addressed is the desirability of distinguishing between the 'theory of' urban regeneration and the role of 'theory (or theories) in' urban regeneration. Whilst both of these aspects of urban regeneration theory are of value, the present discussion concentrates on the former rather than the latter aspect. Although this distinction between the two aspects of theory may appear to be somewhat artificial, the focus of concern in this section is balanced in the following chapters by the inclusion of a number of individual areas of theory that make specific contributions to particular aspects of urban regeneration. Taken together, this section and the following chapters provide a broad overview of 'theory of' and 'theory in' urban regeneration.

Urban regeneration is by its very nature an interventionist activity. Whilst traditionally many forms of intervention have been state led, the desirability of intervening in order to correct a failure of the market has increasingly become a matter of public-private consensus. Although it is tempting to close the debate here, it is insufficient to suggest that consensus can either emerge or continue to function in the absence of the necessary institutional structures. Creating these institutional structures requires the establishment of a central objective (or objectives) and the introduction of a means of mobilising collective effort in order to manage change in an orderly manner.

One attempt to explain and understand the importance of creating a framework within which new forms of collective effort can be developed and applied has been contributed by the regulation school. This theory is based on 'the concept of successive regimes of accumulation' in which 'each regime develops an accompanying mode of regulation' (Knox, 1995, p. 104). Thus, far from eradicating regulation through a reduction in the scope of state activity, the reality is that new patterns of social, political and economic relations emerge. These new forms of control and intervention have emerged as a response to unexpected challenges. In this changed regime the actors involved in urban management and regeneration are required to establish new methods for arriving at consensus. As Healey has argued, one of the most important features of this mobilisation of collective effort is that it encourages a diversity of discourses 'not merely about content, but about the process through which people seek to debate their concerns' (Healey, 1995, p. 256).

Urban regeneration theory is principally concerned with the institutional and organisational dynamics of the management of urban change. However, these institutional and organisational dimensions of the theory of urban regeneration also display a number of important characteristics and features which help to define the role, content and mode of operation of urban regeneration. Given that urban regeneration as a distinct activity is rooted in practice rather than theory, a high degree of similarity between the features of theory and practice is to be expected. Summarising these features, urban regeneration can be seen as:

- an interventionist activity;
- an activity which straddles the public, private and community sectors;
- an activity which is likely to experience considerable changes in its institutional structures over time in response to changing economic, social, environmental and political circumstances;
- a means of mobilising collective effort and providing the basis for the negotiation of appropriate solutions;
- a means of determining policies and actions designed to improve the condition of urban areas and developing the institutional structures necessary to support the preparation of specific proposals.

These characteristics and features reflect the preceding debate, and are chiefly concerned with the role and the mode of operation of urban regeneration.

The other major element of urban regeneration theory relates to the functioning of the urban system as a whole and to the operation of the economic, social, physical and environmental processes that determine the content of urban regeneration. Robson (1988) has identified four main elements of the processes involved in urban change: industrial restructuring in pursuit of maximising returns; factor constraints including the availability of land and buildings; the real or perceived unattractiveness of urban areas; and the social composition of urban areas. The identification of these elements assists in defining the content of urban regeneration. In addition, this assessment also points to the need to consider the ways in which the diverse elements involved in regeneration can be integrated in order to ensure that actions are mutually supportive.

Integration is a central feature of urban regeneration and this feature helps to distinguish urban regeneration from earlier partial attempts to manage change in urban areas (Lichfield, 1992). The earlier sections of this chapter examined the reasons for the existence of urban regeneration and demonstrated the ways in which the various forces of change have been managed in the past. However, whilst not wishing to suggest that individual sectoral initiatives are unwelcome, it is apparent that, for example, an isolated property-led solution cannot be expected to address the full range of economic, social and environmental problems that are encountered in urban areas. Generating and delivering an integrated and comprehensive solution to the challenges of urban regeneration is a difficult task, but it is well worth the effort involved.

This suggests that the final element of a theory of urban regeneration is that it is a strategic process or, to use Hickling's terminology, it is about managing decisions through the use of strategic choice (Hickling, 1974). Given the wide range of issues involved in the management of urban change, and accepting that many individual actions are likely to be of limited scope and short-term duration, it is essential that urban regeneration should work to a strategic agenda. A system for the strategic management of urban regeneration should place emphasis on the need to create clarity regarding the intended outcomes of regeneration, the provision of a framework within which specific plans and projects can be designed and implemented, establishing and maintaining links between the policy systems involved, identifying the roles and responsibilities of the actors and organisations involved in regeneration, and generating a sense of common purpose and co-operation (Roberts, 1990). Matters of partnership, strategy and the management of urban regeneration are dealt with more fully in Chapters 3 and 11 of the present book.

Why Bother to Regenerate Urban Areas?

The preceding sections of this chapter have discussed a number of causes of the 'urban problem', the theories which have been advanced to explain urban change and the consequences of allowing the outcomes of urban problems to continue unchecked. Many different explanations of the origins and occurrence of urban problems have been advanced over the years, and whilst some of these explanations emphasise the influence of an individual event or a particular policy decision, most analysts have adopted a multi-causal explanation. Likewise, it is unusual for the outcome of a process of urban change to be one-dimensional. As is demonstrated in Chapters 3, 5, 6 and other parts of the present book, most urban problems and challenges affect a wide variety of authorities, local communities, special agencies and private firms.

The Causes and Consequences of Urban Change

An urban change event, such as the closure of a factory, may simply be the final outcome of a trading decision taken in a boardroom thousands of miles away. The directors who took the decision may not know the location of an individual manufacturing facility, and in making the decision they are likely to have given little thought to anything apart from the efficient functioning of their company. This tendency to dislocate decisions from their consequences for a local area has been exacerbated in recent years through the internationalisation of production, and this tendency has implications for the role played by firms in schemes for local economic revival (Curran and Blackburn, 1994). It is also the case that many public policy decisions are made without a full appreciation of their spatial consequences.

A similar sequence of cause and consequence may flow from other pro-
pulsive events determined by forces of a different nature. Crime, physical
blight, social polarisation and many other causes can change forever the
composition and social structure of a community or neighbourhood. Physical decay, changing transportation and accessibility requirements, or the
impossibility of adapting buildings to accommodate new uses, can sweep
away an industrial, warehousing, residential or retail district.

The most important implication of the preceding discussion for policy-
makers and practitioners alike is the difficulty of attempting to identify a
single cause of an 'urban problem'. Because many change events are multi-
causal in origin, they reflect a range of influences which emanate from both
within and outwith a city. Whilst at one level the restructuring of towns and
cities can be viewed as representing the outcome of a global process which
implies the restructuring of 'those other critical contexts within which the
world's households live, including the city and the community' (Feagin and
Smith, 1987, p. 13), at another level it may be the case that the future
survival of an economic activity or a close-knit community is threatened by
a political whim or professional misjudgement (Jacobs, 1961; Boddy, 1992).

Before entering into a more detailed discussion on the key elements of
urban change, it is important to distinguish between the problem of the
inner cities and the broader urban problem, and to emphasise the importance of balancing urban problems against potentials. On the first of these
issues, Peter Hall made the point neatly some years ago: 'we need to take
the widest possible view' and to consider the problem of urban change 'in
the spatial context of the rapidly changing economic and social geography
of contemporary Britain' (Hall, 1981, p. 4). Equally, it is important to avoid
dwelling on problems to the exclusion of the potential for positive change.
Kuklinski (1990) has argued that spatial policy needs two principal goals,
economic efficiency and social equity, and that achieving a balance between these can help to resolve problems by mobilising potential. This
suggests that analyses which focus equally upon the competitive advantage
of urban areas, and especially of the inner city, tend to be more helpful
than models which limit their attention solely to the role of welfare policy
in the resolution of problems (Porter, 1995).

Four major aspects of urban change are considered in the following
section:

- economic transition and employment change;
- social and community issues;
- physical obsolescence and new land and property requirements;
- environmental quality and sustainable development.

Economic Transition and Employment Change

Economic change is not a new phenomenon, nor is there any lack of
analysis or policy prescription with regard to this subject. The fundamental
issue that has to be addressed is that many profound changes have occurred

in the structure, profitability and ownership of economic activities. Hannington, writing of the problem of the distressed areas in 1937, pointed to evidence of the breakdown of the traditional urban economic order 'when the basic industries of the system are plunged into continuous slump' (Hannington, 1937, p. 31).

Echoes of this analysis can be identified in more recent studies where the 'urban problem' is seen as part of a broader process of restructuring in which older urban areas have suffered most due to inherent weaknesses in the structure of their economic base and their inability to adapt to new trading and infrastructural requirements (Robson, 1988). In Hall's analysis of the economic performance of urban areas 'goods-handling' places – dependant upon manufacturing, port functions and a range of traditional service activities – have performed worse than 'information handling' places (Hall, 1987), not only in Britain, but also in other advanced economies.

This identification of the fundamental structural weaknesses evident in the economies of the older urban areas led researchers in the 1980s to investigate a variety of causal factors including the 'urban–rural' shift (Fothergill and Gudgin, 1982) and the 'spatial division of labour' (Massey, 1984). More recently, the top-down analyses of the 1980s have been balanced by locally rooted assessments of other aspects of the difficulties experienced by the urban labour force in gaining access to new economic opportunities. In many cases this is seen to be due to the absence of appropriate skills and experience (McGregor and McConnachie, 1995), resulting in the 'social exclusion' of substantial segments of the labour force.

A landmark project of the late 1960s presented an assessment of urban change in unambiguous terms when it argued the case for an integrated regeneration policy in order to address the economic, social and physical decay evident in the inner city of Liverpool. The Shelter Neighbourhood Action Project (SNAP) report (McConaghy, 1972) built upon earlier partial, often sectoral, analyses and stated the importance of viewing urban economic change within the context of broader economic trends, but it also advocated that solutions should be locally rooted. In addition, the SNAP report noted the need to consider the role of the urban, and especially the inner urban, economy within the wider context of the region and nation and it argued that 'it is absurd to attempt to deal with urban deprivation as something quite separate to progress in the urban unit of which it is a part' (McConaghy, 1972, p. 205).

Social and Community Issues
The preceding discussion of economic transition provides an initial insight into the origins of many of the social problems which have beset urban areas. However, economic change, whilst of major significance, is not the only factor that has dictated the scale and occurrence of social problems in towns and cities. Other influences have also been at work; such influences

reflect the evolution of socio-demographic trends, the adjustment and breakdown of traditional family and community structures, the changing nature and outcomes of urban policy, and the consequences of changing social perceptions and values.

Socio-demographic change in recent decades has seen the movement of population away from older urban areas in general, and from the inner cities in particular. This decentralisation of population has been both planned and unplanned (Lawless, 1989). Some households left the city as a result of the construction of peripheral housing estates, the planned expansion of urban areas beyond the immediate sphere of influence of the city of origin, and the building of new towns. However, the majority of those leaving older urban areas have done so through their decision to move to new areas of private housing. The reasons for such moves are many and complex but, in summary, they include the availability of cheaper and more attractive housing, the search for an improved quality of life and the desire to gain access to a better range of services. In addition, this adjustment in residential preferences also reflects the changing location of employment opportunities.

Although the pull of the suburbs and of the free-standing settlements beyond the boundaries of the older urban areas reflects one aspect of this analysis, push factors have also been of considerable importance. Such factors can be seen in most advanced societies; people have moved away from cities in order 'to escape the noisy, crowded city and find space' (Fowler, 1993, p. 7). Urban areas, and especially inner urban areas, are no longer the preferred residential location of the more affluent, instead cities have increasingly experienced a concentration of the poor and disadvantaged members of society. This exclusionary differentiation (Healey *et al.*, 1995) has exacerbated the problems experienced by many urban dwellers, notwithstanding the success of some of the projects aimed at recolonising the city in an attempt to produce a more balanced society.

One of the causes of the changes which have been described above has been the breakdown of traditional structures of community and kinship. The disappearance of traditional sources of employment, the effects of policies aimed at rehousing urban residents, the impact of infrastructure and commercial property development, the decay of the environment, and the absence of adequate social facilities, have combined to erode the cohesion of many urban communities. With the breakdown of the support provided by the neighbourhood, other problems have emerged which have led to further instability and decline. In this situation new issues arise, including the spatial concentration in the inner cities of non-white immigrants and the urban poor. Race is now a significant factor in many of our urban areas, and it is 'important that those concerned with intervention in urban renewal should be particularly aware of the racial aspects and implications of policy' (Couch, 1990, p. 90). New immigrants and the children of earlier generations of immigrants have added an ethnic dimension to many of the issues confronting urban communities. More importantly, and positively, these new groups have also contributed new resources and sources of potential (Oc, 1995).

A final point of importance in this brief introduction to the social and community aspects of urban policy is the image of the city. In the eyes of many people the city is no longer an attractive place that can provide all the requirements necessary for a civilised way of life. Rather, parts of our towns and cities fulfil 'the same role as the howling wilderness of the sixteenth and seventeenth centuries; a place of base instincts, ugly motives, subterranean fears and unspoken desires, a place which reveals the savage basis of the human condition and the frailty of civilised society' (Short, 1991, pp. 47–8). With a public image of this kind can urban areas ever recover their position as the centrepiece of civilised living? The answer can be seen in a number of experiments in social and community regeneration that are aimed at 'breaking out of this stultifying trap' (Robson and Robson, 1994, p. 91), and in the determination of some urban communities, such as the Eldonians of Liverpool, to resist the negative forces of change and to rebuild from within.

Physical Obsolescence and New Requirements

One of the most obvious manifestations of the 'urban problem' is the physical obsolescence of many parts of our towns and cities. In situ decay, the functional obsolescence of buildings, derelict sites, outdated infrastructure and the changed accessibility requirements of the users of urban areas, combine to present a major task for regeneration. Whilst economic, social and institutional factors can be identified that explain the physical decline of cities, in many cases these factors can also be redirected in order to provide the foundations for regeneration. Such an approach can help to guide physical development in order to ensure that it is appropriate and is likely to initiate area-wide physical, economic, social and environmental restructuring. The establishment of a wider mandate for property-led regeneration would help to ensure that physical action for towns and cities also made a greater contribution to the economic and social well-being of such areas (Turok, 1992).

Physical problems arise due to changes in the requirements of the users of urban land and premises, because of the deterioration of the stock of urban buildings and infrastructure, and as a consequence of market failures in the system of land ownership and control. Although there is some evidence to the contrary, there is often a space constraint on the location of economic activities in the inner areas of many cities. This constrained locational view is supported by research (Fothergill, Kitson and Monk, 1983) and by the many instances of firms leaving the city in search of additional space and lower operational costs. Increasing competition for jobs, together with the influence exerted by the new residential preferences of employees, has resulted in the provision of alternative locations that are often better served by modern infrastructure and which offer lower rents or land values (Balchin and Bull, 1987).

Added to these factors are problems associated with the presence of derelict and contaminated land, the cost of clearing sites and providing new infrastructure, and the difficulty of assembling sites. Whilst solutions to such problems are often technically determined and site-specific, it is important to realise that there is an institutional as well as a physical dimension to the occurrence and persistence of urban physical problems. The absence of an adequate institutional capacity to intervene in the cycle of physical decline has proved to be a major impediment to the regeneration of many urban areas. It was in order to address such issues that the new urban initiatives of the 1980s were introduced, including enterprise zones and urban development corporations; the aim of these innovations was to 'experiment with ways of recasting the regulatory regime' (Healey, 1995, p. 262).

A final point of importance that should be noted in relation to the physical problems of urban areas, is the influence of the planning system. Here the evidence is far from conclusive; in some cases blight and neglect have resulted from over-ambitious planning schemes that have exceeded their capacity for implementation, whilst in other instances planning has acted as an enabling force that has generated positive change. What is clear is that achieving urban regeneration requires far more than traditional land-use planning; it has to encompass a broader strategy of urban management which relates 'investment, physical intervention, social action and strategic planning – to other associated policy fields' (Roberts, Struthers and Sacks, 1993, p. 11).

Environmental Quality and Sustainable Development

The final issue to be discussed in this section is concerned with the environment of urban areas. Many of the factors discussed in the preceding paragraphs have conspired to degrade the urban environment. Whilst dereliction is the most obvious outward sign of the imposition of urbanisation on the natural environment, this is not the chief cause of concern. The very existence of what is increasingly referred to as 'unsustainable urbanisation' indicates the origins and impacts of towns and cities that have been developed in order to serve the goal of economic growth. A city 'draws water, energy and many other resources from distant points leaving an environmental or ecological footprint of its consumption pattern' (Roberts, 1995, p. 230).

In many senses urban areas can be seen to generate environmental costs that are not matched by benefits. These costs include the excessive consumption of energy, the inefficient use of raw materials, the neglect of open space, and the pollution of land, water and the atmosphere. Although the 'muck and brass' philosophy of the past may once have been seen to represent the pathway towards a prosperous city, recent research indicates that attitudes and expectations have changed, and that a successful town of the future is increasingly likely to be judged on its environmental performance and appearance (Ache, Bremm and Kunzmann, 1990). Even traditionally attractive features, such as the ready availability of land or a plentiful supply of labour, may in the longer term prove insufficient to ensure the successful development of urban areas. Set against these weaknesses and costs are the environmental benefits associated with urban areas, including the presence of public transport networks, a threshold of population and economic activity that justifies the development of active waste management, and the existence of substantial areas of serviced brownfield land that can be redeveloped.

The new challenge for urban regeneration is to contribute to the achievement of sustainable development. 'The world's economic system is increasingly an urban one' and this system 'provides the backbone for natural development' (World Commission on Environment and Development, 1987, p. 235). New models are now on offer, including 'ecological modernisation' (Roberts, 1997), and their value has now been emphasised in policy statements published by the European Union and the UK Government.

Urban Policy: Origins and Development

The final section of this chapter offers a brief summary of the origins and evolution of modern urban policy, chiefly in England. As is noted in other chapters in the present text, whilst successive British governments have drawn extensively upon the experience of other countries in developing urban policy, it is possible to identify a distinctive British approach to the attempted resolution of urban problems. This approach reflects the apportionment of roles and responsibilities between central and local government, and between the public, private and voluntary sectors. Although the style and content of urban policy has changed in accord with the characteristics summarised in Table 2.1, a number of elements of policy continuity can be identified, including a continuing concern to raise the level of education and training of the urban workforce (see Chapter 7), the need constantly to renew and revise the physical fabric of towns and cities, and the continuing importance of financial and legal factors (see Chapter 9) in determining what can be achieved.

The Early Days

The origins of modern urban policy can be traced back to the 1930s and the designation of slum clearance areas, and to the Comprehensive Development Areas designated under the 1947 Town and Country Planning Act. Further additions to policy occurred during the 1950s and 1960s, including the provisions of Section 11 of the 1966 Local Government Act for special aid in areas of concentration of Commonwealth immigrants and the Educational Priority Areas scheme, which followed the recommendations of the

1967 Plowden Report on primary education (Hall, 1981). In response to growing concern about the condition of the inner urban areas, and especially those neighbourhoods with significant concentrations of immigrants, the Urban Programme was launched by the Home Office in 1968. In 1969 the Local Government Grants (Social Need) Act provided the basis for financial assistance through the Urban Programme.

Other policy initiatives followed, including the Community Development Projects (established by the Home Office in 1969), an expansion of the Educational Priority Areas Scheme and the pioneering SNAP, which was published in 1972. In the early 1970s a series of Inner Area Studies were carried out by consultants and this, together with other initiatives, such as the designation of Housing Action Areas by the 1974 Housing Act, provided the basis for the upgrading of the urban agenda through the 1978 Inner Urban Areas Act. Even though the initial impact of the measures introduced under the 1978 Act was limited to a few inner city areas (Donnison and Soto, 1980), the most important consequence of this legislation was that it placed urban policy in the mainstream of central government policy.

During the 1970s central government responsibility for urban policy changed hands. The control and direction exercised through the Home Office, with industrial and regional policy still within the portfolio of the Department of Trade and Industry, switched in 1975 to the Department of the Environment. This shift in departmental responsibility reflected a change in the emphasis of policy. The Home Office had adopted a social pathology approach, whereas the Department of the Environment emphasised the need for a structural or economic view of urban deprivation and policy (Balchin and Bull, 1987). Under the 1978 Act local authorities could be designated as partnership or programme authorities. A total of seven partnerships, 15 programme authorities and 14 other districts were designated. In Scotland responsibility for urban regeneration was vested in the Scottish Development Agency (established in 1976), which made substantial investments in a number of major area schemes, including the Glasgow Eastern Area Renewal (GEAR) project in Glasgow.

Introducing the Market

Following the change of government in 1979, the Urban Programme continued, but with an increased emphasis on private investment and a greater concern for 'value for money'. Public investment in the Urban Programme increased throughout the early 1980s alongside the introduction of new measures designed to revive and enhance private sector confidence. The first of these new initiatives was the establishment of Urban Development Corporations (UDCs) under the 1980 Local Government Planning and Land Act; two UDCs were designated in 1981, one for London Docklands and one for Merseyside. The second new measure, which was announced in the 1980 Budget speech, was the establishment of Enterprise Zones (EZs); 11 EZs were designated in 1981. Both of these programmes were later expanded; in total 13 UDCs were designated (12 in England and one in Wales), whilst a further 14 EZs were designated in a second round during 1983–84.

Aware that the UDCs and EZs would not be able to address all of the problems of the inner urban areas, the Urban Development Grant (UDG) was introduced in 1982 alongside the establishment of Inner City Enterprises (a property development company, funded in part by the Urban Programme, that was to seek out development opportunities that would otherwise be ignored or considered too risky). Although UDG was in part based on the experience of the Urban Development Action Grant in the USA, there were clear links between UDG and earlier urban policy actions, including the development of special schemes under the Category A Derelict Land Grant arrangements (Jacobs, 1985).

Other urban regeneration initiatives launched during the early and mid-1980s included:

- the establishment of five civil service task forces in the partnership areas (these City Action Teams brought together officials from various central government departments, the Manpower Services Commission, and managers (seconded from industry) charged with the task of unblocking the provision of public services and increasing efficiency;
- the creation of registers of unused and underused land owned by public bodies – this requirement was placed on local authorities by the 1980 Local Government Act;
- the operation and expansion of the Priority Estates Project which was renamed Estate Action in 1987.

In 1987 the Urban Regeneration Grant (URG) was introduced with the intention of complementing the UDG; the purpose of URG was to assist the private sector in bringing forward major schemes. The URG was merged with the UDG into the new City Grant that was introduced in 1988 as the major policy instrument under the Action for Cities Programme. City Grant applications were appraised by private sector secondees and the grant was awarded to developers directly rather than through a local authority intermediary.

Into the 1990s

City Challenge was introduced in May 1991. It invited local authorities to bid for funds in partnership with other public sector, private and voluntary bodies. Eleven bids were selected under the first round (for the period 1992–97) and a further 20 bids were approved in the second (and final) round in 1992. By this stage City Challenge represented the largest single element of the urban policy budget (Mawson *et al.*, 1995). The third round of bidding was suspended and eventually abandoned pending a major review of urban policy.

The outcome of the policy review of the early 1990s was a further move down the road of the 'new localism' (managerial, competitive and corporatist) described by Stewart (1994). What emerged in November 1993 was the Single Regeneration Budget (SRB). The ten new integrated offices in the English regions (the Government Offices for the Regions, or GORs) were given the role of administering the existing main programmes (cuts in budget for both the Urban Programme and the UDCs had been announced in late 1992) and the new SRB. In the November statement the government also announced the introduction of City Pride – a pilot programme which invited multi-agency groups in Birmingham, London and Manchester to develop a ten-year strategic vision for their city and to present an action programme to achieve the vision.

Draft Bidding Guidance for SRB was issued in January 1994, and it indicated that it was expected that most SRB partnerships would be led by local authorities or Training and Enterprise Councils (TECs), although other leadership arrangements were not precluded. Final bidding guidance was published prior to the first bidding round, which commenced in April 1994. In December 1994 the successful 201 bids were announced. The successful projects commenced during the 1995/96 financial year. A further round of SRB took place in 1995 (*Bidding Guidance* – Department of the Environment, 1995 – was issued in March).

The final elements of urban policy introduced during the early and mid-1990s that deserve comment here include the creation of English Partnerships and the Private Finance Initiative. In July 1992 a consultation paper was published proposing the establishment of an Urban Regeneration Agency. The intention was to create a new statutory agency to promote the reclamation and development of derelict, vacant and underused land and buildings in England, especially in urban areas. The agency (English Partnerships) came into full effect in April 1994 and merged the functions previously discharged by English Estates, City Grant and Derelict Land Grant.

In 1992 the government launched the Private Finance Initiative (PFI). The purpose in creating PFI was to reduce the Public Sector Borrowing Requirement and to raise additional capital finance in an attempt to persuade the private sector to take a more active role in urban (and regional) regeneration.

The Current Situation and Beyond

In May 1997 a Labour administration was elected to office. Whilst certain elements of urban policy have rolled forward (the SRB, for example, will continue but with a greater emphasis on the distribution of funds to a wider range of local authority areas), new elements of policy have been introduced. Welfare to Work is a major new employment stimulation measure, alongside the creation of Regional Development Agencies (RDAs).

The newly established Department of the Environment, Transport and the Regions issued, in June 1997, a consultation paper on the proposed RDAs. This indicated the intention to create organisations that will co-ordinate regional economic development, help attract inward investment and support the small business sector. An agency will be created in each GOR region (Merseyside will be merged with the North West Region) and it is anticipated that the RDA will bring together the functions of the GORs with those currently discharged by English Partnerships and the Rural Development Commission.

Beyond the present proposals, and the immediate remit of this section, are other initiatives, including the establishment of a Scottish Parliament, a Welsh Assembly and representative chambers in English regions. These bodies will have an important role to play in the structure and operation of urban (and regional) regeneration.

Other Initiatives and Policies

Three other aspects of policy are of particular relevance to urban regeneration, although they either relate to matters beyond the strict boundary of a discussion of urban policy, or are policy areas outwith the sole determination of UK government. The first issue relates to the continued operation of a wide range of public policies – on health, social policy, housing, education, training, transport, law and order, planning, and environmental standards – that are relevant to urban regeneration. As can be seen from the preceding discussion, whilst some of these policies have addressed urban problems directly, other policies can work against the objectives of urban regeneration. The second aspect of policy, which has existed since the 1930s, is Regional Selective Assistance (RSA). Although the objectives of RSA, and other forms of assistance have varied over the years, are generally coincident with those of urban regeneration, responsibility for RSA is vested in the Department of Trade and Industry and the funds available are used to support projects in urban and rural areas. A final strand of policy, which has grown in importance during recent years, is the European Union's Structural Funds. Many areas of the UK, both urban and rural, are designated as eligible to receive assistance from the Structural Funds. The funds are managed through partnership structures representing the European Commission, the Member State government and local and regional interests (Roberts and Hart, 1996).

Concluding Remarks

This chapter is different to most of the others in this book in that it provides a framework and backdrop, rather than presenting a specific topic.

Therefore, the concluding section of the chapter is more concerned with the identification of matters of interest, than it is with the development of answers or solutions.

However, three general conclusions can be drawn from the preceding discussion. The first is the importance of evaluation in informing the development and further enhancement of regeneration theory and practice. This matter is considered in greater depth in Chapter 10. Second, it is essential to tackle the task of regeneration through the adoption of an integrated and comprehensive approach. Third, it is important to accept that today's new regeneration initiative is but a staging post in the evolution of towns, cities and regions. Regeneration is a constant challenge, and the approach adopted at a particular point in time represents the outcome of a complex system of social, economic and political choice.

The remaining chapters of this book address many of the complexities of regeneration. They also provide insights and guidance that will be of assistance in ensuring that the present generation of contributions to the progress of towns, cities and regions will be both positive and lasting.

Key Issues and Actions

- It is essential to set a context for any proposed regeneration action – this context should consider the historic evolution of an area and the outcomes of previous policies.
- All towns, cities and regions display a particular blend of problems and potentials – this blend is the manifestation of both external influences and internal characteristics.
- The style of approach to regeneration has evolved over the years, and policy and practice reflect dominant socio-political attitudes.
- The regeneration of urban areas can be seen as an important element of regional and national success.
- Urban regeneration is a comprehensive and integrated vision and action which leads to the resolution of urban problems and which seeks to bring about a lasting change in the economic, physical, social and environmental condition of an area that has been subject to change.

References

Ache, P., Bremm, H.J. and Kunzmann, K. (1990) *The Single European Market: Possible Impacts on Spatial Structures of the Federal Republic of Germany*, IRPUD, University of Dortmund, Dortmund.

Balchin, P.N. and Bull, G.H. (1987) *Regional and Urban Economics*, Harper & Row, London.

Boddy, T. (1992) Underground and overhead: building the analogous city, in M. Sorkin (ed.) *Variations on a Theme Park*, Hill and Wang, New York.

Browne, H. (1974) *Joseph Chamberlain, Radical and Imperialist*, Longman, London.

Couch, C. (1990) *Urban Renewal Theory and Practice,* Macmillan, Basingstoke.

Curran, J. and Blackburn, R. (1994) *Small Firms and Local Economic Networks*, Paul Chapman, London.

Department of the Environment (1995) *Bidding Guidance: A Guide to Bidding for Resources from the Government's Single Regeneration Budget Challenge Fund*, Department of the Environment, London.

Donnison, D. (1993) Agenda for the future, in C. McConnell (ed.) *Trickle Down or Bubble Up?* Community Development Foundation, London.

Donnison, D. and Soto, P. (1980) *The Good City*, Heinemann, London.

Fainstein, S.S. (1994) *The City Builders*, Basil Blackwell, Oxford.

Feagin, J.R. and Smith, M.P. (1987) Cities and the new international division of labor: an overview, in M.P. Smith and J.R. Feagin (eds.) *The Capitalist City*, Basil Blackwell, Oxford.

Fothergill, S. and Gudgin, G. (1982) *Unequal Growth*, Heinemann, London.

Fothergill, S., Kitson, M. and Monk, S. (1983) *Industrial Land Availability in Cities, Towns and Rural Areas*, Industrial Location Research Project, Working Paper No. 6, Department of Land Economy, University of Cambridge, Cambridge.

Fowler, E.P. (1993) *Building Cities That Work*, McGill and Queens' University Press, Montreal and Kingston.

Hall, P. (1974) *Urban and Regional Planning*, Pelican Books, Harmondsworth.

Hall, P. (ed.) (1981) *The Inner City in Context*, Heinemann, London.

Hall, P. (1987) The anatomy of job creation: nations, regions and cities in the 1960s and 1970s, *Regional Studies*, Vol. 21, no. 2, pp. 95–106.

Hall, P. (1988) *Cities of Tomorrow*, Basil Blackwell, Oxford.

Hannington, W. (1937) *The Problem of the Distressed Areas*, Victor Gollancz, London.

Hausner, V.A. (1993) The future of urban development, *Royal Society of Arts Journal*, Vol. 141, no. 5441, pp. 523–33.

Healey, P. (1995) Discourses of integration: making frameworks for democratic planning, in P. Healey, S. Cameron, S. Davoudi, S. Graham and A. Madani-Pour (eds.) *Managing Cities: The New Urban Context*, John Wiley & Sons, Chichester.

Healey, P., Cameron, S., Davoudi, S., Graham, S. and Madani-Pour, A. (1995). Introduction: The city – crisis, change and innovation, in P. Healey, S. Cameron, S. Davoudi, S. Graham and A. Madani-Pour (eds.) *Managing Cities: The New Urban Context*, John Wiley & Sons, Chichester.

Hickling, A. (1974) *Managing Decisions: The Strategic Choice Approach*, MANTEC Publications, Rugby.

Howard, E. (1902) *Garden Cities of Tomorrow*, Swan Sonnenschein, London.

Jacobs, J. (1961) *The Death and Life of Great American Cities*, Vintage Books, New York.

Jacobs, J. (1985) UDG: the Urban Development Grant, *Policy and Politics*, Vol. 13, no. 2, pp. 191–9.

Knox, P. (1995) *Urban Social Geography: An Introduction*, Longman, Harlow.

Kuklinski, A. (1990) *Efficiency versus Equality: Old Dilemmas and New Approaches in Regional Policy*, Regional and Industrial Policy Research Series No. 8, University of Strathclyde, Glasgow.

Lawless, P. (1989) *Britain's Inner Cities*, Paul Chapman, London.

Lichfield, D. (1992) *Urban Regeneration for the 1990s*, London Planning Advisory Committee, London.

Massey, D. (1984) *Spatial Divisions of Labour*, Macmillan, London.

Mawson, J., Beazley, M., Burfitt, A., Collinge, C., Hall S., Loftman, P., Nevin, B., Srbljanin, A. and Tilson, B. (1995) *The Single Regeneration Budget: The Stocktake*, Centre for Urban and Regional Studies, University of Birmingham, Birmingham.

McConaghy, D. (1972) *SNAP: Another Chance for Cities*, Shelter, London.

McGregor, A. and McConnachie, M. (1995) Social exclusion, urban regeneration and economic reintegration, *Urban Studies*, Vol. 32, no. 10, pp. 1587–600.

Ministry of Town and Country Planning (1947) *The Redevelopment of Central Areas*, HMSO, London.

Mumford, L. (1940) *The Culture of Cities*, Secker & Warburg, London.

Oc, T. (1995) Urban policy and ethnic minorities, in S. Trench and T. Oc (eds.) *Current Issues in Planning*, Avebury, Aldershot.

Porter, M. (1995) The Competitive Advantage of the Inner City, *Harvard Business Review*, Vol. 73, no. 3, pp. 55–71.

Roberts, P. (1990) *Strategic Vision and the Management of the UK Land Resource*, Stage II Report, Strategic Planning Society, London.

Roberts, P. (1995) *Environmentally Sustainable Business: A Local and Regional Perspective*, Paul Chapman, London.

Roberts, P. (1997) Sustainable development strategies for regional development in Europe: an ecological modernisation approach, *Regional Contact*, NO. XI, pp. 92–104.

Roberts, P. and Hart, J. (1996) *Regional Strategy and Partnership in European Programmes*, Joseph Rowntree Foundation, York.

Roberts, P., Struthers, T. and Sacks, J. (eds.) (1993) *Managing the Metropolis*, Avebury, Aldershot.

Robson, B. (1988) *Those Inner Cities*, Clarendon Press, Oxford.

Robson, B. and Robson, G. (1994) Forward with faith, *Town and Country Planning*, Vol. 63, no. 3, pp. 91–3.

Short, J.R. (1991) *Imagined Country: Society, Culture, Environment*, Routledge, London.

Smailes, A.E. (1953) *The Geography of Towns*, Hutchinson, London.

Stegman, M. A. (1995) Recent US urban change and policy initiatives, *Urban Studies*, Vol. 32, no. 10, pp. 1601–7.

Stewart, M. (1994) Between Whitehall and town hall, *Policy and Politics*, Vol. 22, no. 2, pp. 133–45.

Stöhr, W (1989) Regional policy at the crossroads: an overview, in L. Albrechts, F. Moulaert, P. Roberts and E. Swyngedlouw (eds.) *Regional Policy at the Crossroads: European Perspectives*, Jessica Kingsley, London.

Turok, I. (1987) Continuity, change and contradiction in urban policy, in D. Donnison and A. Middleton (eds.) *Regenerating the Inner City*, Routledge & Kegan Paul, London.

Turok, I. (1992) Property-led urban regeneration: panacea or placebo? *Environment and Planning A*, Vol. 24, no. 3, pp. 361–79.

World Commission on Environment and Development (1987) *Our Common Future*, Oxford University Press, Oxford.

3 Strategy and Partnership in Urban Regeneration

Andrew Carter

Introduction

There is an emerging consensus in Europe, and increasingly in the UK, that in order to address the interconnected problems facing many urban areas there is a need to develop strategic frameworks at the urban region level (Healey, 1997). This consensus is based on the premise that successful urban regeneration requires a strategically designed, locally based, multi-sector, multi-agency partnership approach.

The emergence of such partnerships can be seen as a particular response and challenge to the rapid and fundamental social, economic and institutional changes that society has witnessed over the past few decades. The globalisation and restructuring of the economy have increased the economic, social and physical problems that many cities face, whilst reducing the control that institutions, public and private, have over the economic decisions that affect communities' well-being (Parkinson, 1996). A major consequence of these developments is that the economic fortunes of cities and regions now depend increasingly on the success of local activity. In trying to respond to these changes, we have witnessed a wide variety of policy responses by cities.

Single-sector, single-agency approaches have been proven to have major limitations in trying to tackle the social, economic and physical problems found in many urban areas. 'Gone are the quick fix schemes of the early 1980s. In the place of opportunism and an obsession with getting things done, there is a model of integrated development based on a comprehensive, multi-agency approach' (Roberts, 1997, p. 4). Most organisations involved in urban regeneration (regardless of the needs they are addressing) recognise that the issues they face have multiple causes and therefore need a multi-agency approach to devising and implementing solutions.

This chapter examines a number of major issues:

- the need for a strategic approach to urban regeneration;
- developing a strategic vision and framework;
- principles of a strategic framework;
- the partnership approach in urban regeneration;

- models and types of partnership;
- managing the partnership process;
- policy and practical principles of partnership;
- the strengths and weaknesses of the partnership approach;
- conclusions and future considerations.

The Need for Strategy

The strategic context for urban regeneration has not been well developed in the past. A feature of much urban policy has been a lack of strategic vision and longer-term perspective. The overwhelming emphasis on small areas, discrete projects and output-related funding has left little room for broader considerations (Turok and Shutt, 1994). There has been little or no attempt to devise a strategic view of what should happen to cities as a whole or to individual conurbations. Moreover, it is apparent that many urban policies developed by central government in the 1980s have specifically pursued *ad hoc* projects without attempting to locate these within a broader vision of what should be happening to regions.

> As a result, problems are being addressed in a piecemeal manner and the linkages between different aspects of regeneration have not been developed. Planning and action on a city-wide or regional level have also been sidelined by the focus on local initiatives. Consequently, a duplication of effort is occurring, economic activity is shifted around at public expense and problems of dereliction and deprivation continually reappear and deepen as economic restructuring proceeds.   (Turok and Shutt, 1994, p. 212)

The need for a strategic approach to urban regeneration arises from the concerns regarding property-led urban regeneration and inner city policies in general which have been described as being modest in scale, geographically dispersed, marginal and *ad hoc* in character, and lacking any relationship to structural urban economic trends (Hausner, 1993).

Healey claims that it is 'no longer possible to approach urban regeneration through the promotion of urban transformation projects in isolation'. Instead, she states 'the emphasis should be creating the conditions for economic, social and environmental regeneration' (Healey, 1997, p. 109). Essential in achieving this is the existence of a long-term strategic framework which reflects a process capable of fostering links between issues and those involved in them.

A strategic framework at the urban region level, enables policy parameters to be explored and integrated. Such an examination assists urban regeneration and helps to define the extent to which such measures can in turn meet environmental and social objectives without compromising economic development in the long term.

Box 3.1   Coventry and Warwickshire Partnerships Ltd

The formation of Coventry and Warwickshire Partnerships (CWP) in 1994 was an attempt to provide a platform for strategic economic development for the subregion of Coventry and Warwickshire. The partnership comprises all the area's seven local authorities, the TEC, the Chamber of Commerce, two universities, colleges, private firms, voluntary organisations and trade unions.

The partnership was primarily used to engage the private sector and higher education institutions in the future development of the region. At a strategic level CWP has worked well, in terms of bringing together the key agencies and in securing UK regeneration and European funding. However, at the operational level there has been on occasion conflict between the partnership secretariat and partners' agencies, mainly because of the problem in separating out responsibilities with the partnership.

Source: ECOTEC (1997).

Peter Hall in his review of regeneration policies for peripheral housing estates states that UK urban policy has been characterised by 'inward-looking regeneration policies'. He argues that this represents an unbalanced approach to regeneration. Such policies have failed to tackle many of the root causes of decline. Hall suggests that policy needs to be reoriented towards 'outward-looking policies' (Hall, 1997 p 873), the key characteristics of which are set out in Table 3.1. Such policies should seek to address urban decline through adducing factors in the external environment. This approach emphasises strategic linkages between local initiatives and partnerships, particularly at a region-wide level. This approach also locates particular areas within the wider context of a vision for the urban-region as a whole.

Table 3.1   **Outward-looking policies**

Policy Aspect	Policy Focus
Institutional arrangements	Emphasis on region-wide partnerships; emphasis on horizontal and vertical linkages within and between institutions.
Spatial scale	Linkages between areas of deprivation and potential; region-wide strategic planning frameworks.
Economic development	Education, recruitment and placement; linking local to city and regional development; attracting inward investment.
Social cohesion	Measures aimed at overcoming stigmatisation and social exclusion.
Environment, access and amenity	Overcome physical isolation of declining areas; transport planning; improved amenity to attract outsiders.
Housing	Improve housing to attract new residents; attention to region-wide housing allocation processes.

Source: After Hall (1997).

The European Dimension

According to Alden and Boland (1996) the clearest expression of the European Commission's policy on regional development of spatial planning is to be found in *Europe 2000+*. The document emphasises the emergence of a European dimension in planning policies of member states and advocates an enhanced role for regional development strategies in achieving the objectives of nations, regions and localities in an enlarged European Union (EU).

Within this context, *Europe 2000+* identifies a number of major trends emerging from member states which advocate the role for spatial planning in devising and implementing regional development strategies.

The first major trend is the growing awareness that spatial planning has shifted from a concern for purely physical planning and land-use matters to a wider concern for social, economic, environmental and political issues. This reflects the return of the importance of strategic thinking in planning, especially given the importance of cities and regions operating within both the EU and even wider global economy where strategic levels of decision-making assume significance (Alden and Boland, 1996).

A second trend examined in *Europe 2000+* is the need not only to identify strategic issues, but also to integrate the various issues into a more comprehensive and complex form of spatial planning. Planning at a variety of spatial levels is now concerned with a wider range of issues than hitherto, including economic development, transportation, retailing, tourism, housing, urban regeneration, the countryside, and their integration with each other.

Third is the increased decentralisation of responsibility for policies and controls to regional and local levels of government. This is coupled with an increased number of organisations responsible for delivering services within the region.

The ability to think and act strategically is essential for the long-term success of the UK in Europe. As Roberts argues, 'In the absence of strategic vision it is doubtful if the UK will be able to sustain a range of viable economic, social and physical environments within a Europe of regions' (Roberts, 1990, p. 6). The benefits of applying the principles of strategic vision are already apparent in a number of European countries; the strategic vision approach has been used to help plan and regenerate a number of cities and regions, including Barcelona.

Elements of a Strategic Approach

There is a growing consensus amongst policy-makers about what the elements of a strategic approach to regeneration might be. As Parkinson (1996) notes, when the European Commission revised its Structural Funds in 1988 it identified four necessary features of the reformed policy: it should bring added value, be partnership based, be clearly targeted and integrate different policy instruments and approaches. Similarly, when the Scottish Office introduced its New Life for Urban Scotland initiative in 1988 it defined the strategic approach to regeneration as comprehensive, multi-sectoral and partnership based. This thinking finds echoes in more recent UK central government regeneration initiatives – City Challenge and the SRB.

In trying to define the elements of a strategic approach to regeneration, McGregor *et al.* suggest that it is concerned with the use of resources to secure lasting social and economic change by making complementary investments in interacting local activities. These changes are designed to facilitate further desired change and to have beneficial effects on other sectors and areas within the local or urban economy. This approach implies some knowledge of how investments and other changes interact to promote local change, to create a dynamic and to produce positive spill-overs (McGregor *et al.*, 1992).

Parkinson (1996) drawing on this work and other research states that a strategic approach to urban regeneration should:

● have a clearly articulated vision and strategy;
● specify how its chosen mechanisms and resources would help to achieve the long-term vision;
● clearly integrate the different economic, environmental and social priorities of the regeneration strategy;
● identify the intended beneficiaries of the strategy and the ways in which they will benefit;
● identify the level of private, public and community resources, financial and in kind, that would be committed over defined periods of time;
● specify the role and contribution that the public, private and community partners would make to regeneration;
● integrate, vertically and horizontally, the policies, activities and resources of those partners in a comprehensive strategy;
● link explicit regeneration policies to wider mainstream programmes in housing, education, transport, health, finance which constitute the implicit urban strategy;
● specify the relationship between short, medium and long-term goals;
● establish economic, social and physical baseline conditions before the policy intervention to allow an assessment of change over time;
● have agreed milestones of progress;
● monitor the outputs and outcomes of the strategy and evaluate their impact.

As Parkinson rightly acknowledges, this represents a formidable set of criteria which in the real world of policy would be extremely difficult to achieve. Despite this it is worth stating them as an ideal typical set of criteria, against which to evaluate actual strategy development.

Principles of a Strategic Framework

Devising a strategic framework requires skills in 'making links – setting contexts to foster relationships; and, strategic vision – mobilising ideas about the future' (Healey *et al.*, 1995, p. 284). This framework should:

- provide a bridge between 'top-down' and 'bottom-up' approaches;
- be realistic and capable of being translated into specific policies, objectives and actions;
- be drawn up by a wide-ranging partnership, which includes all key stakeholders;
- address the overall viability, prosperity and competitiveness of regions – enhancing their contribution to their own residents and to their regions and the nation;
- ameliorate disadvantage, promote opportunity and mobility, support development in deprived communities;
- preserve cities as motors of civilisation, culture, innovation, opportunity and enterprise (Hausner, 1993).

Strategic vision is concerned with creating the framework in which longer-term goals, aims and objectives of individuals, organisations and areas can be realised (Roberts, 1990). The creation of a strategic vision can ensure that resources, for example land, capital and labour, are used in such a way as to achieve the best overall effect. A strategic vision should emphasise:

- the interdependence of actions, rather than treating each action as independent;
- the long-term outcomes and benefits, other than mainly considering short-term costs;
- the overall requirements of an area, rather than stressing the potential of an individual site (or project);
- the importance of creating common ground and, wherever possible, the generation of consensus, rather than encouraging conflict;
- the creation of positive attitudes towards mutual collaboration between sectors, rather than maintaining a public-private sector divide.

There are important lessons to be gained from the experience of attempts to devise and implement strategic vision:

- It is important to be aware of the complexities that are involved in constructing an approach to (resource) management which is based upon strategic vision.
- There is a need for consistency of purpose, the benefits of adopting strategic vision are unlikely to be fully evident in the short term.
- It is important to encourage the widest possible participation of both 'bottom-up' and 'top-down' interests in the setting of goals, in generat-

ing a vision or visions, in identifying and obtaining the necessary resources and in the management of implementation.

- It is desirable to create a system for strategic vision and management
  which is self-sustaining and which recognises, at the outset, that there
  will be a need to adjust and fine-tune the policies which are pursued.
- It is equally important to ensure that, having created consensus and
  agreed upon a strategic vision, the processes of implementation also
  adhere to agreed objectives.
- It is desirable to monitor, to evaluate regularly and disseminate information widely on the progress of the agreed strategy.

Strategic planning is an important tool for enabling communities to identify advantages in relation to the external environment – local, regional,
national and international. This emphasis on external factors allows the
process to incorporate a wide range of organisations and individuals from the
public, private, voluntary and community sectors. Clearly the partnership
approach is a critical element in adopting a strategic approach to urban
regeneration. This approach is explored in greater detail in the next section.

The Partnership Approach

The ethos of partnership, and multi-agency provision and collaboration,
have become the key concepts of the 1990s. Although seven Inner City
Partnerships were formed under the 1978 Inner Urban Areas Act, the
approach only really developed significantly during the late 1980s. By the
early 1990s there was a consensus developing between all the main political
parties that a closer involvement between the public and the private sectors, together with the direct participation of local communities and the
ability to cut across traditional policy boundaries, were all essential elements of an effective urban regeneration strategy (Bailey, 1995). This
principle has since been extended to practically all aspects of public policy
– training, housing, community care and social services.

Despite the recognition of the need for partnership, which unites different levels of government and other public, private and community
actors and agencies, 'the problem of generating the right institutional ma-
chinery with adequate incentives, sanctions and resources to integrate the
actions of national and local, of public, private and community institutions
and agencies – to make partnership a reality rather than a cliché remains a
challenge' (Parkinson, 1996, p. 31).

Why Partnership?

There are a number of main reasons behind the move towards multi-
agency partnerships as the preferred method of working in addressing a

wide range of social, economic and environmental issues. First, the current political agenda is forcing the pace in this area. Funding requirements for initiatives, such as City Challenge and the SRB, explicitly require the development of partnerships. The Guide to Bidding for resources from the government's Single Regeneration Budget states that, 'bids must be supported by partnerships representing an appropriate range of interests which should include relevant interests in the private and public sectors and in local voluntary and community organisations' (Department of the Environment, 1995, p. 2).

Second, the multidimensional and complex nature of urban problems requires integrated, co-ordinated and multifaceted strategies involving a wide range of actors. The concerns raised throughout the 1980s and the 1990s regarding both property-led urban regeneration and inner city policies point to the need for a longer-term, more strategic, integrated and sustainable approach to urban regeneration, which incorporates a broader package of programmes for finance, education, business development and social provision. Partnerships are perceived to be the most effective vehicle for achieving these goals. Advocates of partnerships argue that, because they offer greater involvement by all sectors in the decision-making process, they are seen to be an inherently more efficient and equitable way of allocating public funds.

Third are the difficulties associated with the centralisation of power and fragmentation of duties and organisations involved in urban areas. Partnerships which involve a wide range of agencies and organisations can help to co-ordinate activity and extend across traditional policy boundaries.

Fourth, in many policy spheres, for example, housing, education, health care and crime prevention, individuals are challenging the paternalistic nature of central and local government initiatives. Local people are increasingly demanding a voice in defining and implementing the most appropriate responses to the challenges facing their locality.

Models of Partnership

There is no single model of partnership. Mackintosh puts forward a useful framework for understanding the process of partnership. Partnership is, she argues, a concept in public policy which 'contains a very high level of ambiguity' (Mackintosh, 1992, p. 210) with its potential range of meanings subject to 'conflict and negotiation'. Mackintosh devises three main conceptual models of partnership in relation to the urban regeneration context:

- The synergy model suggests that by combining their knowledge, resources, approaches and operational cultures, the partner organisations will be able to achieve more together than they would by working on their own or, in other words, the whole is greater than the sum of the parts.

- The budget enlargement model is based upon the knowledge that by working together the partners will gain access to additional funds that neither could access on their own.
- The transformational model (with a different focus) suggests that there are benefits to be gained by exposing the different partners to the assumptions and working methods of other partners (that is, it will stimulate innovation as part of a continuing process of development and change) and Mackintosh suggests that successful partnerships always result in such transformation.

In the current context, policy-makers have been particularly concerned with partnerships based upon the first two models. A key aim has been to achieve more with the same inputs or, increasingly, with less. To be more cost-effective while finding new ways of gaining access to additional resources. The Bidding Guidance states that 'bids must show how Challenge Fund resources will reinforce other public sector initiatives, maximise the leverage of additional private sector investment'. It is also envisaged that bids will 'harness the talents and resources of the voluntary sector and volunteers and involve local communities' (Department of the Environment, 1995, p. 3).

Types of Partnership

Over the last ten years, the variety of partnerships has expanded significantly. Partnerships are overlaid upon complex organisational and political environments and the number of potential partners in any one area is large and diverse. The nature of partnership organisations makes categorisation difficult. The strategies deployed vary according to local circumstances, national and local policy, and the interplay of the different interests within the partnership arrangements (Bailey, 1995).

Table 3.2 offers a typology of alternative categories of partnership. This typology seeks to define six categories, with examples taken from contemporary urban regeneration in the UK. It is not structured in such a way as to evaluate a broad range of partnerships, but it illustrates the variety that exists and points to the capacity of those responsible for urban regeneration to explore new working relationships among different and sometimes competing groups in the economy.

Partnership describes both an organisational structure, bringing together a range of agencies to co-operate to achieve shared objectives, and a structure for policy-making. Partnerships can operate at different levels:

- Systematic partnership involving strategic policy-makers are most effective in dealing with large-scale, deep-rooted problems.
- Programmatic partnerships might tackle issues such as the implementation of an urban regeneration strategy.

Table 3.2 **Types of partnership**

Type	Area of coverage	Range of partners	Activities	Examples
Development partnership, joint venture	Single site or small area, e.g. town centre.	Private developer, housing association, local authority.	Commercial/non-profit development producing mutual benefits.	London Road Development Agency, Brighton Media Centre
Development trust	Clearly defined area for regeneration, e.g. neighbourhood or estate.	Community-based, importantly, independent from public bodies but frequently some reps from Local Authorities.	Community-based regeneration. Generally concerned with creating and spreading community benefits and since they are non-profit making, recycle all surpluses into the trust.	Coin St, North Kensington Amenity Trust, Arts Factory
Informal arrangement	District or city-wide.	Private sector-led. Sponsored by Chamber of Commerce or development agency.	Place-marketing, promotion of growth and investment. Concerned with problems, issues and strategy identification that are of mutual interest to the parties involved.	The Newcastle Initiative, Glasgow Action, East London Partnership.
Agency	Urban, or subregional.	Terms of reference from sponsoring agency. Delivery may be through a team of secondees drawn from the partners or through a development company (limited by guarantee) that is independent of the partners.	Multiple task orientation, usually within a designated time-frame.	UDCs, City Challenge, SRB, 'New Life' Partnerships.
Strategic	Subregional, metropolitan.	All sectors.	Determining broad strategy for growth and development. May act as an initial catalyst for activity. Often acts as a guide for development. Implementation is often through third parties. Can act as an umbrella organisation guiding other vehicles, including development companies.	Coventry and Warwickshire Partnership, Chester City Partnership, Thames Gateway London Partnership.

Source: After Bailey (1995) and Boyle (1993).

- Technical partnerships may be short-term arrangements to achieve a particular objective such as a discrete physical redevelopment project.

To understand both the differences and the similarities between partnership arrangements, Stewart and Snape (1995) have identified three 'ideal types' or organisational models of partnership; these are illustrated in Box 3.2. Each reflects a different understanding of three key dimensions to partnership – the nature of partnership objectives; the relationship between partners; and the specific activities of partnerships.

Box 3.2   Models of partnership

Facilitating partnerships	Negotiation of contentious or politically sensitive issues; partners have differing perspectives; wide-ranging objectives; focus on deep-rooted problems; powerful stakeholders; balance of power crucial.
Co-ordinating partnerships	Drawing together partners to oversee initiatives undertaken by the partners themselves or by arms-length bodies; address relatively new and non-contentious issues; often led or managed by one partner; balance of power not as delicate.
Implementing partnerships	Specific objectives and time limited; responsible for the delivery of agreed projects often involving securing funding and resources; outputs clearly defined; power relations unproblematic.

Source: Stewart and Snape (1995), p. 4.

Many partnerships contain elements of all three models, whilst in others there remain doubts about the appropriate balance between facilitation, co-ordination and implementation.

Managing the Partnership Process

Research undertaken for the Joseph Rowntree Foundation by the Ashridge Centre (in 1997) highlights the key issues that need to be considered in developing a successful partnership management process (see Box 3.3).

Principles of Partnership

Whilst the scale and scope of the partnership, and the type and number of actors who should be involved, will vary according to the aims and objectives established, there are none the less certain defining principles which should underpin the process.

The quality of the partnerships that are formed is of critical importance. Research suggests a symbiotic relationship between the quality of the partnership and the quality of the regeneration strategy (Carley, 1995). 'Successful partnerships reveal a capacity to adapt to changing conditions: political, economic and commercial. They demonstrate the "loose-tight" characteristics of successful organisations: clearly pursuing well-determined strategic objectives while retaining the ability to adapt tactically to overcome impediments and obstacles' (Boyle, 1993, p. 321). The most robust partnerships are those which respect the roles and contributions of each of the partners; the most productive are those which are flexible and reflective; and the most beneficial are those which are sustainable beyond the requirements of a specific programme.

Box 3.3   Managing the partnership process

Stage 1
Partners come together through mutual recognition of a common need, or in joint effort to obtain public funds. If they have not worked together before, they begin the process of overcoming differences in background and approach, building trust and respect. There may be the need for training, building each partner's capacity to operate effectively in this new organisation.

Stage 2
Through a process of dialogue and discussion, the partners establish common ground and work towards agreeing a vision and mission statement for the initiative. The original core group of partners might agree on the need to involve more individuals and organisations in the initiative. The partners develop mechanisms for assessing needs and quantifying the size of the task they propose to undertake. The initiative combines the information generated by the needs assessment exercise, together with the vision and mission statement to produce an agenda for action.

Stage 3
The formal framework and organisational structure of the partnership is designed and put in place. The partners set specific goals, targets and objectives linked into the agenda for action. Where appropriate, the executive arm of the partnership selects or appoints a management team to oversee the work of the initiative.

Stage 4
The partnership delivers to its action plan, whether through service provision or some other function. The executive arm seeks to maintain the involvement of all partners, formulates policy decisions and ensure the continuing process of assessing, evaluation and refining the operations of the partnership.

Stage 5
Where appropriate, the partners should plan their forward strategy. This involves developing a new set of goals for the survival and continuation of the work of the initiative in some form. They should seek to create 'life after death' by transferring the asset of the partnership back into the community with which they work.

Source: Wilson and Charlton (1997).

Partnerships must be built on shared interests, reciprocal support and mutual benefit with each partner contributing according to their respective resources, strengths and areas of expertise. The varying requirements of each partner, such as the need for public accountability of governments, profit for private sector organisations and personal gratification for volunteers, must be recognised.

Box 3.1   Thames Gateway London Partnership

The Thames Gateway London Partnership (TGLP) consists of 12 local authorities, two TECs, the London Docklands Development Corporation (LDDC) and English Partnerships. The origins of TGLP lie in a report carried out by SERPLAN (London and South East Regional Planning Conference) in the early 1990s which looked at redressing the imbalance between East and West London. The result was the Thames Gateway Planning Framework. The TGLP does not have any strategic power, it is a strategic partnership not a strategic authority. All decisions are made by consensus.

The TGLP has developed a vision for the area which is encapsulated in the planning framework. The challenge lies in turning the vision into reality. Whilst the vision is based on long-term objectives of 20–30 years, the partnership recognises the need for early success in order to keep the partnership together.

Stages of Development

Boyle (1993) in his analysis of partnership working in West Central Scotland has identified a number of key stages through which partnerships progress:

(1) Launch and need for early wins to establish credibility.
(2) Implementation of early action programmes.
(3) Consolidation and reassessment of aims and objectives.
(4) Longer-term, more ambitious programmes of structural change.

As partnerships proceed through these stages, and in many instances before they actually reach Stage 1, a number of policy and practice issues emerge which have important implications for their development and ultimately their success. The following set of issues are an attempt to provide a provisional 'check-list' of issues to consider in forming partnerships. This is clearly not an exhaustive list and as the experience of partnership development and working grows and deepens this list will be expanded.

Policy Issues

Strategic Context
Recognising the external context and fundamental socio-economic realities of all partnerships is vital. 'It is essential that the underlying condition of

the regional and national economy, the state of the local labour market, the broader political context and the harsh realities of the commercial world are brought to the attention of the partnership' (Boyle, 1993, p. 322). Partnership objectives need to be realistic and attainable, given the powers and resources available.

Individual schemes function best when clear priorities are set within the overall context of a strategic plan, recognising what can be realistically achieved within the resources available. A strategic framework can provide the setting within which a multiplicity of neighbourhood regeneration initiatives can establish their local visions of how they contribute to the locality as a whole. This approach allows the key strengths and weaknesses of an area to be analysed and individual area initiatives to be planned and managed within the overarching strategy.

Box 3.5   Chester Action Partnership

The partnership is a multi-sectoral network established in 1993 with a programme to highlight and address the challenges facing Chester City and its rural district. It currently represents more than 300 organisations and leads the development and implementation of both new and existing strategies for regeneration.

Its role is to facilitate action and provide a strategic framework for all activity taking place in Chester. The programme represents a coherent set of goals, priorities and actions which has provided a framework in which individual projects can be implemented as the appropriate resources are secured. Current activities include: managing £10 million SRB Programme; managing Estate Action Schemes; City Centre Management; 'Living Over the Shop' Project; 'Clean Up' Campaign; Cultural Strategy; Anti-Poverty Strategy; government and private sector secondments and sponsorship; 'Arts for Health' project.

Integration

From an organisational perspective, Carley (1996, p. 8) identifies both 'vertical' and 'horizontal' integration as prerequisites for sustainable regeneration. Vertical integration is the beneficial linkage and co-ordinated policy and action at appropriate spatial levels: national, regional, local, neighbourhood and the household. This is 'subsidiarity' to use the EU term, in which action at each level is important, but not sufficient on its own for the achievement of policy objectives.

Horizontal integration has two aspects. First, cross-sectoral connections between departments in central and local government to generate more efficient responses to multiple deprivation and the requirements for sustainable regeneration. Second, the bringing to bear of all appropriate stakeholders, or interest groups, in partnership to address the complex challenges of multiple deprivation.

Engaging Key Actors

A vital stage in the establishment of a partnership is to identify stakeholder groups across all sectors and take appropriate action to ensure they do not feel alienated and threatened. More effective and accountable partnerships have equal representation of all the relevant actors. As well as identifying key stakeholders to be involved, there is a need to ensure they have access to information, management procedures and decision-making powers.

Focus of Activity

Partnerships often need to demonstrate quick successes in order to create confidence and support in local communities. Partnerships use tried and tested procedures, notably housing and environmental improvements, to legitimise their existence in difficult and controversial situations. There is a danger that more ambitious initial objectives involving, for example, innovative business development or community development, can be marginalized by the demands on the partnership to deliver. Partnerships often face real political and organisational difficulties in moving from low-risk, traditional projects to more creative, risky ventures (Boyle, 1993). Long-term external support is necessary to temper the immediate demands of the partnership with the basic realities of the situation and longer-term benefits that may derive from slower, steady progress.

Building Networks

Successful partnerships depend heavily on the operation and quality of networks (Skelcher, McCabe and Lowndes, 1996). Effective networks give separate agencies added strength through combining with others. In particular, they offer the possibility of greater information exchange and the development of a shared perspective among diverse groups. Networking involving those affected by an initiative increases its potential to be more sensitive to the needs of the locality.

Practice Issues

Leadership

If a partnership does not have the leadership capacity and creative skills to engender a common sense of purpose and develop a shared vision, then it is unlikely to use regeneration resources well. Political will and support are prerequisites of effectiveness while the active involvement from senior actors in participating agencies can demonstrate real commitment. Without basic support, the energies of partnerships are easily subsumed by internal conflict and constant struggles to secure additional resources. In this context it is crucial that multiple agencies are involved early in the process and that partners in different sectors are mutually aware of each other's strategies. In addition, it is also essential to ensure that the chief executive (officer) and the convenor are able to work together and that they command the support of the partners.

Independence

Management also needs to be independent of a single sector interest and representative of the full range of stakeholders. Partnership arrangements seem to work best where staff are appointed or seconded to an independent partnership agency rather than being the responsibility of an existing agency. Likewise, partnerships require their own resource budget in order to exert more influence on other partners through negotiation, project selection and leverage.

Staff

The quality of the personnel involved appears to be a critical factor for achieving success over the longer period. Their ability to develop meaningful relationships with politicians and representatives from the participating agencies is often a vital ingredient of effectiveness. There is also evidence of the transfer of experience, with key officials taking their expertise from one initiative to another. This serves to highlight the importance of bureaucrats in the evolution of partnership and raises the possibility that professional/technical concerns can overtake, or even replace, the original problems identified by the community or its representatives (Boyle, 1993).

Measuring Success

Monitoring and evaluation are central to the ability of partnerships to demonstrate that they have achieved worthwhile results. The measurement of partnership achievements needs to be further refined and better tailored to partnership objectives. The pressure for objective quantifiable measurement, the absence of satisfactory qualitative measures of changed attitudes, raised awareness, improved image, capacity building etc., and the difficult measurement of 'visionary' objectives have all emerged as key issues which have to be resolved (Stewart and Snape, 1995).

Creating Added Value

A further element in partnership evaluation which needs to be addressed is whether the partnership provides added value. This involves the assessment of additionality (whether resources applied through partnership achieve more than would have resulted from their individual use elsewhere), synergy (whether two or more partners energise each other to make two and two equal five), and displacement (whether partnership activity has meant the loss of some activity elsewhere in the locality).

The Challenge of the Partnership Approach

There is widespread acceptance of the need for comprehensive approaches to deal with the complex problems of contemporary urban areas and to advance strategic objectives for cities as a whole. In addition, there appears to be a growing realisation that some flexibility in the construction of partnerships is essential at the local level and that it is the primary task of central government to provide an effective policy framework without undue interference at the local level. However, as Bailey notes 'much remains to be done to create the necessary linkages between locally based partnerships, central government budgets and, of increasing importance, access to European structural funds' (Bailey, 1995, p. 226).

Strengths of Partnerships

Partnerships have the potential to ensure that the weaknesses of previous approaches to urban regeneration are overcome by concentrating upon those elements which have a maximum impact upon urban problems. In addition, partnerships can bring together social, economic and physical activities within the same strategy. They can bring a new dynamism to old problems and are often effective in forging new links between existing stakeholders.

Box 3.6   Mansfield Diamond Partnership

The partnership provides a comprehensive approach to the problems caused by the pit closures in Nottinghamshire. The desire for strong community involvement provided the initial focus for this strategic partnership. The partnership co ordinates over 80 projects in a three-year £50 million SRB programme in the Mansfield area of Nottinghamshire. The term 'diamond' aptly conjures an image of the many facets involved including business development, crime and safety, community development, education and training and general environmental improvements. The scheme demonstrates the catalytic effect that government funding has had in developing closer links between organisations located in the area.

Partnerships can create synergy between programme and policy areas such that impacts can be achieved which are beyond the reach of any one stakeholder, for example by accessing and integrating new sources of finance (Mackintosh, 1992).

Improved co-ordination and delivery of local services can be achieved through creating a setting within which local priorities can be identified. Involving community representatives is particularly important here, as well as establishing efficient communication channels between partner organisations. Where good working relationships are established, it is possible to break down interdisciplinary barriers and to develop comprehensive programmes.

The weaknesses of local individual partners can be overcome by joint action, while their strengths are consolidated by the partnership. Private sector partners provide business acumen and are likely to become more socially responsible, providing they can continue to make effective business decisions. Community organisations are flexible, close to informal networks and support the long-term interests of the community; but they often have few financial resources. Governments, both local and national, have key roles to play in encouraging co-operation, supporting local initiatives and decision-making, leveraging resources and providing administrative support (Bailey, 1995).

Things to avoid

Avoiding the 'lowest common denominator' approach to partnerships is needed to encourage bolder actions and to turn talking-shops into proactive, problem-orientated ventures, with more equal sharing of risk among partners.

Initial enthusiasm to set up the partnership, often to a tight time-scale, means that important management, representation and accountability issues, and reporting arrangements are overlooked or sidelined, leading to conflict and stagnation later (Stewart and Snape, 1995).

There is the tendency for the lead agency to create partnership in its own image (Mawson *et al.*, 1995). The private sector tend to opt for lean, small agencies managed by leading corporate executives, whereas local authorities tend to create large bureaucratic organisations. In community-based partnerships there is often little private sector representation and the reverse occurs in private sector-led examples. This can lead to a lack of balance in determining priorities.

Conflicts can arise between different levels of commitment and involvement of different partners. Perceptions vary as to the uneven distribution of costs and benefits and the lack of time in the early stages often means that 'capacity-building' is overlooked.

Some partners are difficult to engage. Where partnerships are unstable or where goals are not shared, stakeholders are motivated mainly by the need to protect their vested interests. Variations in power and political clout, in annual budgeting, in planning cycles and work styles (especially language) put different pressures on partners and can create tensions within the partnerships (Parkinson, 1996). Considerable time and effort therefore needs to be spent in building partnerships within an atmosphere of trust and sharing. Participation and partnership-working requires a new culture of joint action if they are to deliver sustainable results.

Conclusions

The potential benefits of adopting a strategic approach to urban regeneration are readily apparent. It encourages local authorities and other organisations to establish clear aims, in the context of an agreed strategy, offering them the opportunity to devise criteria against which the merits of individual projects can be assessed. This makes the selection of schemes which reflect longer-term aspirations much more likely (Roberts, 1990).

The actual development of a strategic framework can help to foster collaboration and encourage partnerships between the broad range of agencies involved in urban regeneration. Moreover, it can be employed as a basis for monitoring and evaluating the outcomes of urban policy initiatives, thus enabling a more informed approach to future policy formulation, resource allocation and project appraisal (Martin and Pearce, 1995).

A strategic vision and framework needs to emphasise:

- a genuine working multi-sectoral partnership;
- the co-ordination and integration of initiatives, rather than focusing on single issues;
- a long-term commitment, rather than stressing short-term outputs and costs;
- the development of local regeneration strategies.

As Martin and Pearce note:

> At the heart of this process is the need to develop projects and programmes which are consistent with agreed policies and to identify and implement projects which respond to programmatic need. The interrelationships between policies, programmes and projects would therefore, need to be expressed clearly, and projects selected on the basis that their objectives were consistent with those defined at both policy and programme levels. (Martin and Pearce, 1995, p. 109)

There now exists a considerable body of knowledge on the organisation and management of partnership-based strategic vision. Yet, in creating such frameworks there is clearly no universal approach. Hence, there are no ideal solutions because the choice of approach must depend not only on the results desired, but also on the availability of partners, the gravity and nature of the problems, the existing organisational and institutional structures and the national ethos.

Partnership-working on the whole has been evolutionary, pragmatic and piecemeal with individual people and organisations 'making it up as they go along' (Stewart and Snape, 1995, p. 11). There has been little exchange of experience about the process of managing partnerships. Individuals and partnership organisations are seldom asked to provide feedback about their experiences and lessons learnt.

Many of those involved in partnership-working, especially organisations set up prior to SRB, are now in a position to develop 'good practice' guidelines both for the development of new partnerships and for the conduct of existing partnership activities. Some partners are already developing internal guidance of this type, but it is in the interest of effective partnership-working that all parties potentially involved in regeneration partnerships should recognise basic ground rules for partnership-working.

Drawing on the Organisation for Economic Co-operation and Development's (OECD's) (1996) and BURA's (1996; 1997) experience as to what constitutes good practice in urban regeneration, it is possible to suggest a number of principles for policy in relation to the establishment of successful partnerships:

- A strategic vision and framework, providing a clear picture of the desired outcomes, encourages partners to align their goals and objectives while making appropriate contributions. Partnerships should be built on shared interests, joint understanding and action.
- Partnerships should be developed to suit local and regional conditions. The specific characteristics of the locality, the stakeholders and vested interests should determine the structure, composition and mode of operation of the partnership.
- Partnerships should combine both 'bottom-up' and 'top-down' initiatives. Capacity-building and mutual understanding are essential across all the sectors to ensure that the partnership can work effectively.
- Partnerships cannot work in isolation. Locally based projects need to be integrated into a wider framework for the region. They require the support of agencies at the level of the city, region and nation, if their actions are to be successful and sustainable.
- Effective partnership working requires clear allocation of responsibility within partner organisations, accompanied by adequate resources, time and structures.
- Partnerships should involve local residents and community organisations as equal partners. This often requires a change in culture and way of operating to accommodate community participants. The involvement of these groups is necessary to ensure their full commitment to achieving the jointly established goals and to ensure that they are the principal beneficiaries of whatever action is taken.

This chapter has demonstrated the considerable role that partnerships, both strategic and local, can play in devising and implementing urban regeneration initiatives. It also suggests that there is a degree of consensus among policy-makers and practitioners about the gains which can be achieved from effective partnership-working, and about some of its costs and current limitations.

Mike Geddes identifies two primary gains from working in partnership:

- Partnerships can provide a strategic framework which encourages the co-ordination and integration of policies and resources between the public, private and voluntary sectors and with local communities themselves.
- This can enable social and economic policies to respond more effectively to local characteristics, and introduce a greater degree of

differentiation and responsiveness to the needs of local stakeholders and social groups (Geddes, 1997).

He concludes by stating that, 'there is no one model of partnership which can be advanced as the optimum for all contexts. There is a need for continuing flexibility, innovation and experimentation, supported by research and evaluation, to advance good practice' (Geddes, 1997, p. 130).

Key Issues and Actions

- Partnerships should be built on a clear strategic vision and framework for action.
- Partnerships should reflect shared ownership, shared interests, common ambitions and joint understanding.
- Partnerships must be tailored to local conditions and should aim to be inclusive.
- Within partnerships a clear allocation of responsibilities is essential.
- Partnerships will change over time.

References

Alden, J. and Boland, P. (1996) *Regional Development Strategies: A European Perspective*, Jessica Kingsley, London.

Bailey, N. (1995) *Partnership Agencies in British Urban Policy*, UCL Press, London.

Boyle, R. (1993) Changing partners: the experience of urban economic policy in west central England, 1980–90, *Urban Studies*, Vol. 30, no. 2, pp. 309–324.

British Urban Regeneration Association (BURA) (1996) Best practice awards 1996. Booklet prepared for the BURA Best Practice Awards Ceremony, 21 May, Sheffield.

British Urban Regeneration Association (BURA) (1997) Best practice awards 1997. Booklet prepared for the BURA Best Practice Awards Ceremony, 3 June, London.

Carley, M. (1996) Sustainable development, integration and area regeneration. Paper prepared for the Area Regeneration Programme for the Joseph Rowntree Foundation.

Carley, M. (1995) *Using information for sustainable urban regeneration*, Scottish Homes Innovation Paper 4, Edinburgh.

Department of the Environment (1995) *Bidding Guidance: A Guide to Bidding for Resources from the Government's Single Regeneration Budget Challenge Fund*, Department of the Environment, London.

ECOTEC (1997) *Planning in Partnership: A Guide for Planners*. A Final Report to the Royal Town Planning Institute, RTPI, London.

European Commission (1994) *Europe 2000+*. Cooperation for European territorial development, European Commission, Luxembourg.

Geddes, M. (1997) *Partnership against Poverty and Exclusion?* The Policy Press, Bristol.

Hall, P. (1997) Regeneration policies for peripheral housing estates: inward and outward-looking approaches, *Urban Studies*, Vol. 34, no. 5–6, pp. 873–90.

Hausner, V. (1993) The future of urban development, *Royal Society of Arts Journal*, Vol. 141, no. 5441, pp. 523–33.

Healey, P. (1997) A strategic approach to sustainable urban regeneration, *Journal of Property Development*, Vol. 1, no. 3, pp. 105–10.

Healey, P., Cameron, S., Davoudi, S., Graham, S. and Madani-Pour, A. (eds.) (1995) *Managing Cities: The New Urban Context*, John Wiley & Sons, Chichester,

Mackintosh, M. (1992) Partnership: issues of policy and negotiation, *Local Economy*, Vol. 3, no. 7, pp. 210–24.

Martin, S. and Pearce, G. (1995) The evaluation of urban policy project appraisal in R. Hambleton and H. Thomas (eds.) *Urban Policy Evaluation: Challenge and Change*, Paul Chapman, London.

Mawson, J., Beazley, M., Collinge, C., Hall, S., Loftman, P., Nevin, B., Srbljanin, A. and Tilson, B. (1995) *The Single Regeneration Budget: The Stocktake Interim Report Summary*, University of Birmingham and Central England, Birmingham.

McGregor, A., Maclennan, D., Donnison, D., Gemmell, B. and MacArthur, A. (1992) *A Review and Critical Evaluation of Strategic Approaches to Urban Regeneration*, Scottish Homes, Edinburgh.

Organisation for Economic Co-operation and Development (OECD) (1996) *Strategies for Housing and Social Integration in Cities*, OECD, Paris.

Parkinson, M. (1996) Strategic approaches for area regeneration: a review and a research agenda. Paper prepared for the Area Regeneration Programme for the Joseph Rowntree Foundation.

Roberts, P. (1990) *Strategic Vision and the Management of the UK Land Resource*, Stage II Report, Strategic Planning Society, London.

Roberts, P. (1997) Opinion, *BURA Journal*, October.

Skelcher, C., McCabe, A. and Lowndes, V. (1996) *Community Networks in Urban Regeneration: 'It all depends who you know!'*, The Policy Press, Bristol.

Stewart, M. and Snape, D. (1995) Keeping up the momentum: partnership working in Bristol and the West. Unpublished Report from the School for Policy Studies to the Bristol Chamber of Commerce and Initiative.

Turok, I. and Shutt, J. (1994) The challenge for urban policy, *Local Economy*, Vol. 9, no. 3, pp. 211–15.

Wilson, A. and Charlton, K. (1997) *Making Partnerships Work: A Practical Guide for the Public, Private, Voluntary and Community Sectors*, Joseph Rowntree Foundation, York.

PART 2

MAJOR THEMES AND TOPICS

4: Economic Regeneration and Funding

David Noon, James Smith-Canham,
Martin Eagland

Introduction

Economic regeneration is a vital part of the process of urban regeneration. It is needed to counter the economic decline experienced by cities with the changes in the workings of the economy and the increasing globalisation of markets. The industrial revolution made cities the focus of production, population, culture and society in Britain but the last three decades have been characterised by a loss of employment and population, a relocation of manufacturing and the readjustment of other economic structures. After a long period of continued growth the advent of decline led to questions being asked about the role of the city in a modern economy, including the need for cities to readjust to these processes of urban and regional change which have seen the economic benefits of a city location decline in favour of urban fringe and more rural locations. The development of urban policy over the past three decades has been a process of responding to the continually changing nature of a modern economy and to its spatial manifestations.

This chapter considers a number of major issues:

- definition and scope of economic regeneration;
- the workings of urban economies;
- early attempts at economic regeneration;
- economic regeneration initiatives and programmes;
- the economic impact of regeneration;
- financing economic regeneration;
- future considerations.

Components of Economic Regeneration

The successful economic regeneration of a city is two-sided, both demand and supply processes are at work.

Demand Side

The demand side will be determined by the city's ability to retain local expenditure and to attract more spending from outside. This may take the form of demand for industrial production or for the output of the service sector. Efforts to promote the city aim to attract new sources of expenditure. For example, the development of conferences and tourism markets have been pursued as part of strategies to sustain new types of economic activity in the city.

Supply Side

On the supply side investment must be made to improve infrastructure including the building of new roads or improving existing roads and other communication links. Land needs to be redeveloped and made available to both existing industries, keen to expand or relocate, and also for the development of new industries. Important links between education and research institutions have a role to play in the development and operation of science, technology and business parks. There must also be investment in people with increased and appropriate training and support for ideas delivered through enterprise and training bodies such as the TECs and local enterprise companies (LECs), and business support agencies such as Business Link.

During the Industrial Revolution cities were the focus of production and population with close proximity to activities generating external economies which have been termed 'spatial economies of scale'. The economy is now changing, spatial economies of scale are of less importance, and the cities need to react and plan for the future. The economy is entering a new phase where the handling and communication of information is becoming increasingly important. Directly linked to this is the increasing use of information and communication technologies, often referred to as telematics. Cities, to ensure their future role in the changing economic climate, need to have excellent telecommunication networks, not just internally or with other urban centres in Europe but with the rest of the world. Without the physical establishment of these networks and their optimal utilisation, an urban area risks moving from being a core regional or national location to a peripheral global location. The supply side will also be influenced by the productive capacity of the city. Of particular importance here is the ability to attract investment and for indigenous development to be promoted through new firm formation and through the growth of existing businesses. The climate for investment in productive capacity will clearly be influenced by the competitiveness of the local economy as reflected in the quality of infrastructure, skills of the workforce and locational advantages.

The 1996 White Paper *Competitiveness: Creating the Enterprise Centre of Europe* focuses on these supply side issues and sets out policy aims for developing the information society, raising the role and profile of small and medium-sized enterprises (SMEs), encouraging further deregulation, and ensuring a suitably trained, skilled and well-qualified workforce. These initiatives give emphasis to the view that spatial economic performance is the result of the key factors of production, land, labour and capital being the major areas for public policy intervention when seeking to address urban and regional economic disparities.

To succeed, economic regeneration initiatives therefore need to address both the supply-side and demand-side components. There is little point in providing major new infrastructure without sufficient demand to sustain its use, whilst the promotion of a town and city in the absence of adequate infrastructure and facilities is destined for failure.

Theory of Urban Decline

It has been generally recognised that there has been a need for urban regeneration in Britain from the 1960s when it started to become increasingly clear that many cities were facing long-term decline. This decline was characterised particularly by population and employment loss with a net out-migration of population, firms and activities. Linked to these two major factors has been physical and social decline.

During the 1960s much of the literature on the workings of the urban economy was centred on explaining the growth of cities. Central place analysis and urban base theory were both popular approaches to explaining urban growth and change (see Richardson, 1971, for a review). These were demand-oriented approaches. The supply-oriented theories of urban growth stressed the role of population growth, capital investment and technical progress and particularly the attraction of these growth factors from outside the city. These theories were relevant when considering the continued growth of urban centres. However, it was clear from the mid-1960s that growth in many centres had stopped and cities were mainly facing decline.

Several explanations have been advanced for urban decline, and Table 4.1 below aims to summarise the main economic interpretations and explanations which have been put forward.

Explanations of urban economic decline are often based entirely on empirical evidence but they do have an important role to play in urban regeneration. If explanations can be found for decline then lessons can be learned and past mistakes avoided. For example counter-urbanisation is caused by factors such as people's desire to move to the countryside because of the negative aspects of living in a city such as congestion, pollution and crime. In conjunction with some people's desire to relocate firms move out of urban areas because of the push factors of the city such as high service costs, development constraints, failing infrastructure and

Table 4.1   **Explanations for urban decline**

Name of explanation	Explanation
Structuralist	Structural change in the global economy; rise of new economies with differing spatial/locational requirements.
Counter-urbanisation	Pull factors of rural areas and the push factors arising in urban areas causing firms and population to move out.
Marxist	Need to maximise exploitative potential of capital by using cheaper, flexible, less militant labour that can be found in less urbanised areas.
Sectoral or planning	Unintentional effects of spatial planning policies such as development of greenbelt encouraging firms to move well away from urban centres.
External ownership	Increasing external ownership of firms in urban areas by others with little local allegiance.
Product-cycle	Standardisation means that manufacturing can happen almost anywhere so production elsewhere becomes highly probable.

increasingly inflexible labour. To prevent this urban regeneration strategies have focused on improving urban infrastructure, reducing the reclamation and assembly costs of urban land through a range of grants and enhancing the skill base of the local workforce through targeted training schemes designed to improve the job prospects of the local workforce.

Economic Regeneration Programmes

Early Attempts

Early attempts at economic regeneration programmes are exemplified by Boxes 4.1 and 4.2, case studies of Glasgow, Lowell (near Boston, USA) and Dean Clough (Halifax). These examples are characterised by a comprehensive approach to economic regeneration which served to bring together a number of separate initiatives or by the vision and commitment of key people.

Trends in Urban Economic Policy

From the mid-1960s to the present day it is possible to identify four phases in the development of urban policy as a response to urban decline. The first phase ran from the mid-1960s until the publishing of the 1977 White Paper (HMSO, 1977). During this time there was a particular view of deprivation based on the 'culture of poverty' thesis, developed in America by writers such as Banfield (1970), which generally attributed urban problems to the

---

Box 4.1   Glasgow

In the late 1960s and early 1970s Glasgow was suffering an economic crisis as markets for steel, shipbuilding, railway engineering and other engineering activity slipped away. The industrial base was in serious decline and unemployment was rapidly rising. Since then, Glasgow's fortunes have been successfully reversed, its image enhanced and its popularity as a visitor city stands in stark contrast to those earlier depressing images. A number of reasons for the city's success in regenerating itself can be identified:

- the creation of the Scottish Development Agency in the mid-1970s;
- a stable local government;
- the ability to create partnerships between many agencies;
- a strategic approach that gave priority to the creation of new markets for the city rather than a site-specific approach.

This approach gradually brought benefits from visitors attracted to the city and the inward relocation of companies and people which in turn stimulated other investments in the city.

Today, the successes of the 1980s are being consolidated and further developed by the Glasgow Regeneration Agency.

Box 4.2   Lowell, Massachusetts, and Dean Clough, Halifax

The economic fortunes of many towns in the north of England declined with the slow but steady disappearance of their textile industries. From the 1960s onwards many large, and often striking, textile mills stood empty or were underutilised. In the 1970s and 1980s Lowell near Boston in the USA led the way in showing how such mills could be reused and economic life returned to the local area. Their success was based upon a number of factors:

- the presence of high-tech industries and the willingness of some of these industries to occupy the mills after refurbishment;
- local tax incentives for investment;
- capitalising on the presence of local waterways to enhance refurbishment schemes;
- the ease with which local partnerships were created;
- a flexible approach to the application of building regulations.

A similar development in a textile mill is Dean Clough in Halifax where 1 million square feet of space was converted into a number of business and cultural uses through the vision of a local businessman, Sir Ernest Hall. The success of the project was based around flexibility in letting arrangements, introduction of a cultural element, involvement of the local community combined with the attraction of several blue chip tenants.

inadequacies operating within families concentrated in small areas. 'It assumed poverty was a limited problem concentrated in small areas within which a definable anti-social culture could be identified and ultimately eradicated' (Lawless, 1988, p. 532). Therefore, many of the urban economic programmes launched during this period represented an attempt to tackle the 'social' problem. However, the Community Development Projects (CDPs) and the Inner Area Studies (IASs) helped to change this attitude towards deprivation and signalled a new direction in urban policy.

The second phase ran from 1977 through to the early 1980s; the 1977 White Paper signalled the change in focus, and the change in government following the 1979 election saw this phase develop further. This was a period characterised by an emphasis on the development of land and premises to bring about urban economic regeneration. The underlying theoretical approaches were associated with a concern that there were major supply constraints in operation, with limited land for the expansion of existing businesses and few opportunities to attract major new inward investment. It was accepted that major public sector funding would be required to improve inner city development opportunities and to address the problems of dereliction. This was the beginning of the economic approach to urban regeneration with the launch of Industrial Improvement Areas, Derelict Land and Urban Development Grants and Enterprise Zones. Urban policy at this time was starting to show the signs of becoming increasingly centrally focused with the role of local authorities and other local bodies gradually being diluted.

This shift from local to central control continued with the next phase of urban policy from the early 1980s until 1987/88. The government saw the local authorities as overly bureaucratised and inefficient resulting in high local rates which were burdening local businesses. Therefore, the private sector was encouraged to become actively involved in urban regeneration facilitated by new agencies such as the Urban Development Corporations (UDCs). This policy of 'leverage planning' involved the development of mixed use and often larger schemes aided by urban regeneration grants which were made available by central government to local authorities and UDCs. The other theme to be developed during this period was that of partnerships which involved the public and private sectors. Partnerships were encouraged to create a more efficient, dynamic approach to urban regeneration through central government initiatives applied locally such as City Action Teams (CATs) and the Task Forces. The concept of public sector funding levering in private sector investment and the notion of partnership between various agencies and organisations have continued as a central plank of urban policy through to the mid-1990s.

During the 1980s the flagship project was a major focus for economic activity and urban regeneration in the 1980s. It was usually an exceedingly large mixed-use scheme, with a commercial emphasis and bringing with it major infrastructure and land reclamation benefits. Flagship projects sought to stimulate very substantial investment and customer demand while matching this with accommodation on the supply side. In their favour, supporters of flagship projects can point to the successful promotion of the recipient area, generation of external demand and worthwhile improvements in spending power locally. Critics point to the distortion of the local marketplace and the absorption of many years' demand into one part of the area. A number of flagship projects were either killed off or very much reduced in scale by the recession of the late 1980s and early 1990s. However, in more recent times the top slicing of European Regional Development Fund (ERDF) funding for Regional Challenge bids demonstrates the attractiveness to policy-makers of high profile initiatives which can gain wide publicity and acclaim.

Also in the 1980s urban regeneration initiatives had their focus gradually shifted from the local communities and authorities to being controlled and run by central government and the various quangos it established. Following the release of the 1988 *Action for Cities* (HMSO, 1988) document this trend has been reversed and there is now a desire to increase the role that local communities and local authorities play in urban regeneration. In this fourth and final phase of the chronology of urban policy there has been an increasing recognition that the nature of urban problems varied between different localities and that urban regeneration strategies needed to be developed and 'owned' by the local communities. An example of this new approach to urban economic regeneration is provided by the Coalisland Regeneration Project.

Box 4.3　Community Involvement: Coalisland Regeneration Project

The Coalisland Regeneration Project, a partnership of the local community, public agencies and local business, has made significant improvements to the town centre of this declining industrial town by providing a community facility in the refurbished Cornmill, car parking, a water feature and landscaped area, and the conversion of a derelict factory and vacant premises into economic use. The community has been involved as an active partner in the development process, and this has led to a deeper feeling of ownership. The strength of the regeneration lies in the energy of the community which has led to a strong partnership approach.

Individual commercial property owners have been given financial assistance to refurbish vacant buildings and improve shop frontages. The regeneration is the basis for further renewal, and future plans include a canal link, industrial tourism and more small work units. The Coalisland scheme illustrates the powerful leverage effect of public money when well applied.

Programmes such as City Challenge and the Single Regeneration Budget (SRB), together with the establishment of the Government Offices for the Regions, represent a commitment to increasing local participation. This greater local and regional emphasis is also, not least in the European context, a necessary move as it ties in with the EU's regionalist ideals and was an important step in improving the chances of securing funding from this source.

Economic Regeneration Initiatives and Programmes

The previous section has focused on the development of urban policy from the 1960s to the present day. Developments over this period have resulted in the establishment of a range of policy initiatives which are described in this section. The initiatives presented here have formed a significant component of urban policy during the 1990s.

Grant Support

Land grants have been an important element in many urban regeneration schemes such as at Merry Hill in the Black Country and, on a smaller scale, with the redevelopment of the Bird's Custard Factory in Birmingham. Such use of grant support lowers the entry threshold for the private sector and can induce significant investment as a result.

Box 4.4   The Custard Factory, Birmingham

The former Bird's Custard Factory in Digbeth, Birmingham, had been derelict for many years before a City Grant Award of £800,000 was made in January 1992. This public sector funding levered in £1.6 million of private sector investment for the refurbishment of 100,000 sq. ft of redundant buildings, providing 145 units for use by artists, designers, communicators and others associated with creative design.

The development provides low-cost starter units with a range of services designed to support new businesses. The first phase has created some 300 jobs with 50 per cent of occupants being previously out of work. When the second and third phases are completed it is anticipated that about 1,000 jobs will have been created.

The first land grant was the Derelict Land Grant (DLG) which was extended after the Aberfan disaster of 1966 and was initially more applicable to rural areas. During the 1980s the focus of the grant shifted to the treatment of derelict urban areas. Derelict land grants support a proportion of the costs of bringing land back into beneficial use. Local authorities qualified for 100 per cent grant if they were in Assisted Areas or Derelict Land Clearance Areas; 75 per cent in National Parks and 50 per cent elsewhere. For the private sector it was 80 per cent in Assisted Areas and Derelict Land Clearance Areas and 50 per cent elsewhere (DoE, 1988a; 1991). Local authorities were the dominant users of the DLG and take-up by the private sector was relatively low.

Urban Development Grants were established in 1982 to encourage private sector resources to flow into urban development through the application of a minimum public sector contribution. There was to be no restriction on the type of project, although from 1984 onwards the emphasis was clearly on physical regeneration projects where the contribution from the private sector was required to be several times greater than that from the public sector. The grants helped to stimulate additional private sector investment and they created inner city housing in places where it would not have been built. However, only a small proportion of the Urban Programme authorities benefited and these were concentrated in very small pockets of cities. There were also criticisms about the complex and lengthy application procedure.

Urban Regeneration Grants were launched in 1987 to overcome some of the constraints of the UDGs with the main change being that developers could directly approach the government and bypass the local authority. To qualify the sites for development had to be larger than 20 acres. The URG and UDG were replaced by the City Grant – which was close to the model of the URG – outlined in the 1988 *Action for Cities* document. It, like the URG, bypassed local government and was offered by the Department of Environment (DoE) regional offices to offset specific disadvantages of an inner city site and assist the project's commercial viability.

Relaxation of Regulations

Enterprise Zones were established in 1981 based on the principle of removing the 'burden on business' through the relaxation of planning controls and providing tax benefits. Twenty-eight EZs and four extensions have been designated by the government since 1981. Each zone operated for a period of ten years from its initial designation. Some zones, therefore, have already reached the end of their lives. In 1996 five zones were operating, including two extensions. Three further zones are to be designated in areas which have suffered job losses due to the contraction of the mining industry.

The benefits currently available to businesses situated in these zones include:

- exemption from non-domestic rates;
- 100 per cent allowances for corporation and income tax purposes for capital expenditure on industrial and commercial buildings;
- a much simplified planning system and a reduction in government requests for statistical information.

The findings from an interim assessment (HMSO, 1987) on the effectiveness of Enterprise Zones led the government to announce that it did not foresee extending the scheme generally although it did suggest that new zones could be created if they were seen as the best way of tackling a particular local problem. Two such cases did occur, one in Sunderland in 1990 and another in Inverclyde in 1989.

Urban Development Agencies

The passing of the 1980 Local Government Planning and Land Act gave the government the power to create Urban Development Corporations, the first two of which were established in 1981 followed by a further 11 between 1987 and 1993. These were agencies created to aid the regeneration of specific urban localities mainly with the money, ideas and focus of the private sector primed by central government funding. These agencies had the power to acquire, improve and service areas of land and then to act as their own development controller. The basic aim was to ensure that private investment was encouraged to play its part in urban regeneration. By 1989 the UDCs had dealt with 40,000 acres and had a central grant of £200 million per annum. In 1988 the then government argued that the UDCs were 'the most important attack ever made on urban decay' (HMSO, 1988).

English Partnerships, originally known as the Urban Regeneration Agency, was set up through the 1993 Leasehold Reform, Housing and Urban Development Act. It was formed as a type of 'roving UDC' when the third round of City Challenge was cancelled. This new quango reflected the government's continuing emphasis upon land and property development and inward investment, but was set up to develop links with and be committed to local communities.

Integrated Approaches

In the early 1980s concern was expressed about the lack of integration and co-ordination of urban policies and initiatives. In response to the need for a more holistic approach to be applied across a local area and a recognition of the importance of a local community input, City Action Teams were created. This experiment commenced with the creation by the government of the Merseyside Task Force 1981. Following the relative success of this pilot project, six CATs were established in 1985 in the Partnership Areas. Each team consists of the Regional Directors of the Department of Environment, Employment and Trade and Industry and these teams were responsible for the co-ordination of central government expenditure totalling £850 million per annum, instilling private sector confidence, looking for market funding and providing a contact point for external sources of funding for the area.

However, they in themselves did not bring any additional resources to the area. They have also not always worked closely with local governments, which was their aim, and it is also stated (Lawless, 1988) that many of the projects the CATs have initiated or been involved with would probably have happened anyway. The CATs were subsumed within the Integrated Regional Offices in April 1994.

Task Forces were launched by the government in February 1986, initially numbering eight. The aim of the Task Force is to better co-ordinate the efforts of government departments, local government, the private sector and the local community to regenerate inner city areas.

The Task Force consists of a team of about five civil servants. Teams are then supplemented with secondees from local authorities and the private and voluntary sectors. Task Forces are based in some of the most deprived inner cities in the country such as Bradford, Hull, Moss Side and Hulme (Manchester).

Box 4.5   Aims and activities of Task Forces

Main aims of Task Forces

- Increase employment prospects for residents by removing barriers to employment.
- Create and safeguard jobs.
- Improve employability of local people by raising skill levels and providing training.
- Promote local enterprise development through support for enterprise training, financial and managerial assistance.
- Support education initiatives.

Typical activities of Task Forces

- Encouraging enterprise by attracting business through the development of premises and the provision of financial and managerial support.
- Matching work skills with current and future work opportunities.
- Supporting training schemes to improve employability.

By the spring of 1994 Task Forces had committed about £148 million to 5,800 projects. These projects have helped to create over 31,000 jobs, provided over 175,000 training places and helped over 44,000 businesses.

Competitive Bidding for Funding

The 1990s have been marked by the allocation of urban regeneration funding through a process of competitive bidding based on clear objectives, output measures and value for money. An important element has been the involvement of the private sector and the prospects of attracting significant private sector funding.

The City Challenge programme was launched in 1991 and represented a new era of competitive bidding for funding. It differed from many previous policies in that it sought to give the local authority a key role by letting the authority draw up plans for the regeneration of areas that they felt were pivotal in the region's resurgence. Again it involved strong links with the business and commercial sectors but this time it was also to draw upon the resources of the local voluntary sector, which indicated a move away from solely economic regeneration motives to include a stronger social policy dimension.

City Challenge encouraged the local authority to act with vision and to include the local people and community organisations (with, of course, the private sector) in its projects. Stress was placed on an integrated approach to link urban projects with employment and training, childcare, housing, environmental concerns, and crime prevention and safety. The innovative part of the programme was that it introduced a competitive bidding process with the aim of encouraging an entrepreneurial ideal into local government. It was also felt that even those areas that 'lost out' in the bidding process would benefit from the process and work better with their new partners in the private sector.

Reports by the Audit Commission (1989) and the National Audit Office (1990) identified a degree of overlap in the government's inner city programmes and concluded that this amounted to a waste of both time and funds. These reports stimulated the formation of the Single Regeneration Budget in 1994 which was to incorporate the 20 existing programmes and initiatives into one integrated regeneration budget. The SRB introduced a wider and more varied range of regeneration activity than before. Integration and co-ordination on a more local level, initiated in the City Challenge programme, were further encouraged and enhanced through the SRB. This local focus was further supported by the establishment of Government Offices for the Regions in 1993 to establish control of spending and policy implementation at the regional level and to address the shortcomings of past partnership attempts because of a failure to involve local communities.

Experience to date suggests that the SRB is likely to be dominated by projects aimed at large-scale physical development. This is because both the UDCs and English Partnerships Programmes, including land reclamation, are financed from SRB funding (Table 4.2).

Table 4.2   **The Single Regeneration Budget (£ million)**

	1994–95	1995–96	1996–97
HATS	88.2	90.0	90.0
UDCs	291.0	253.7	244.7
English Partnerships	180.8	210.8	220.8
All other SRB programmes	887.1	777.6	768.2

Source: Department of the Environment

The SRB signals a change in the primary influences acting upon urban economic policy. Much government urban policy has been drawn from experiences and ideas from North America, particularly the USA. In the case of the SRB many parallels can be drawn from the French Contrat de

Ville. This initiative encourages partnerships between the state and local institutions in order to form a mutual understanding of the underlying causes of problems and how best to tackle them. The initiative has produced cohesive partnerships at the local level which override administrative boundaries and departmentalism both centrally and locally, and it has achieved an impressive degree of co-ordination of public policy and expenditure (Oakley, 1995a; 1995b).

The SRB now represents the main form of support from central government for urban regeneration in the UK.

Financing Urban Regeneration

Since the start of modern urban programmes following the 1977 White Paper, there have been several variations in the methods used to provide finance.

The main source of funding for urban regeneration in the early days of the programmes was the public sector. Prior to the publication of the 1977 White Paper, the Urban Programme was administered by the Home Office who provided 75 per cent grant aid to many local authorities and also to some voluntary groups. In 1977 the Labour government substantially increased the Urban Programme from £30 million per annum to some £125 million per annum and moved its administration to the Department of the Environment. The seven urban partnerships created in 1978 received £66 million directly from the Urban Programme budget and a further £66 million from a £100 million inner city capital programme.

Following the 1979 general election new ideas emerged on how to fund urban regeneration. The new focus was to encourage greater participation by the private sector in urban regeneration initiatives. This amounted to 'neo-liberalism underwritten by state intervention . . . a subsidy to the private sector through infrastructural investment' (Atkinson and Moon, 1994, p. 165). Of course, there was still a considerable input of direct government funding, the emphasis was shifted towards the private sector with the hope that a degree of entrepreneurial initiative would be instilled by the private sector into the local authorities and other bodies associated with regeneration. This change in policy direction reflected the desire of the Conservative administration for an economic focus to urban policy and it introduced a development-led process concentrating on refurbishment, building, infrastructural improvement and redevelopment.

In a document published in 1983, Michael Heseltine stated that it is 'a major concern to maximise the amount of private sector investment in the inner city areas' (Heseltine, 1983, p. 21). Many of the policies of the Conservative government during this time were based on what can be described as 'leverage planning' (Atkinson and Moon, 1994, p. 192). This is where the main aim of public investment was to 'lever in' private investment.

Leverage planning where 'additional private resources are raised through public expenditure, so as to increase the net benefit of public spending for economic regeneration' (Robson, 1989) can be effectively demonstrated by some of the Conservative programmes introduced in the 1980s, many of them based upon similar initiatives introduced in the USA.

An example of this is the Urban Development Grant which was based on the USA's Urban Development Action Grant. This was a scheme in which local authorities worked up capital investment projects in co-operation with private sector interests. The private sector was to provide much of the investment finance for the project. The scheme was then submitted to the Department of the Environment. There were no restrictions on the type of project eligible for this grant and the scheme involved competitive bidding for a share of the capital resources available. Successful projects were awarded with a grant paid at the rate of 75 per cent of the finance that the local authority put forward for the project.

Another example of the application of the principle of 'levering in' can be seen in its approach to providing low-cost housing. In this case the public sector was invited to bid for derelict land grants which could be used to reclaim land upon which the private sector could then build low-cost housing. In 1983 Michael Heseltine stated that £5 million of public money offered for this scheme attracted £14 million of private sector money, i.e. £19 million of capital housing investment at a cost to the exchequer of only a quarter of the total amount.

The Urban Development Corporations which started in 1980 are often thought to 'represent the flagship of Conservative government urban policy' (Imrie and Thomas, 1993, p. 1) despite the Audit Commission's (1989) concern that the UDCs were over complex and did not have the resources to match the scale of the problem. The UDCs are another property-led attempt at regeneration with the aim of 'levering in' private sector resources. The UDCs receive funding from two different sources: an annual budget from central government, and the utilisation of receipts from the sale of land and property.

The government channelled much of its energy and resources into the UDCs during the 1980s as they became a priority area and a focus for property-led regeneration. Between 1981 and 1990 UDC funding was £1.8 billion, a large proportion of which went on land purchase and assembly (Imrie and Thomas, 1993). The UDCs were successful in 'levering in' large sums of private sector finance. The Public Accounts Committee (1989) noted that the London Docklands Development Corporation levered in over £2 billion in 1989 and the Tyne and Wear Development Corporation had £250 million of private sector funding committed in the first two years of operation.

A major problem faced by the UDCs was the virtual absence of returns during the early 1990s on investments in land and property following the slump in the property market in 1989. Many of the UDCs have recorded losses in this area of activity during the early 1990s.

The focus on the private sector and its role in urban regeneration was reflected in the 1988 Action for Cities programme. The literature for this programme described how the government was spending £3 billion on the inner cities through its various initiatives and it also launched a new consortium called British Urban Development (BUD) which was made up of senior executives from 11 large construction and engineering companies. This consortium had plans to invest in new development schemes. It signalled the start of another partnership where government would facilitate and enterprise, through BUD, which would ensure effective implementation. The slump in the property market killed off this initiative.

As discussed earlier in the chapter the 1990s have witnessed a sea change in the operation of urban regeneration, with the emphasis moving away from grants and quangos to competitive bidding, joint ventures, partnerships and the increasing importance of European finance.

The first widespread example of this new approach was City Challenge. In 1993 the government brought together the Departments of Trade and Industry, Employment, Transport and Environment into Integrated Regional Offices of Central Government (IROs). This was accompanied by a merging of all existing regeneration programmes into a single framework named the Single Regeneration Budget, which is administered by the IROs. The SRB funds all the regeneration quangos (UDCs, the Housing Action Trusts – HATS – and English Partnerships), with any remaining money open for competitive bidding through a Challenge Fund. Bids for SRB challenge funding need to be supported by partnerships with TECs and local authorities expected to play a strong if not leading role within them. This Challenge Fund is not currently spatially targeted (though need must be proven) and as a result has led to a wider geographical spread of regenerative resources, including pockets of deprivation or dereliction in generally affluent areas, for example, the Roundshaw Estate, London Borough of Sutton.

English Partnerships (EP) came into operation in April 1994. The agency focuses on physical regeneration initiatives throughout England but concentrates on Objective 1, 2 and 5b areas, coalfield closure areas, City Challenge and inner city areas, and other assisted areas. Its objective is to promote job creation, inward investment and environmental improvement through reclamation and development of vacant, derelict and underused or contaminated land and buildings – acting in strategic partnerships with local authorities, the private sector, voluntary bodies and others. English Partnerships also has certain statutory powers giving it the ability to provide financial assistance (grants, loans and guarantees), enter into joint ventures, carry out development and purchase land and buildings (through compulsory purchase or by agreement). Every aspect of its involvement in regeneration is financed by its investment fund, which comes from the government's Single Regeneration Budget.

Since 1995 the National Lottery has provided a major new financial resource for a wide range of partnership-based, unique and often striking projects which otherwise would fall outside the scope of conventional public funding or be too demanding for it. There are five 'good causes' supported by Lottery funds. These are the Millennium Commission, the Sports Council, the National Heritage Memorial Fund, the Charities Board, and the Arts Council. These bodies, through the Lottery funds, are supporting both major and minor projects for conserving the national heritage, environmental improvement, community development, sports and recreation, and the arts. Some of the projects supported continue to act as a catalyst for the regeneration of local areas. Major flagship projects supported so far by the Millennium Commission include the Lowry Centre in Salford, the International Centre for Life in Newcastle, Bristol 2000, Millennium Point in Birmingham, and the Earth Centre, Doncaster. A number of similar prestige schemes are currently under consideration.

The European Commission, through its allocation of regional development aid, is a significant provider of funds for urban regeneration in the UK. Many local authorities and local regeneration partnerships gain significant additional resources from the Commission to implement projects which they would otherwise be unable to finance. The vast majority of these come from the Structural Funds, which are intended to promote economic and social cohesion in the EU through redistributing resources in favour of less prosperous regions. The Structural Funds concentrate on four regional policy objectives of which the following three apply to the UK:

- Objective 1, for regions where development is lagging behind the Community average (70 per cent of the total funds available);
- Objective 2, for the adjustment of regions worst affected by industrial decline (11 per cent of the funds);
- Objective 5b, for structural adjustment in rural areas (4 per cent of the funds).

A further 9 per cent of the total resources are spent on Community Initiatives which can also be used to help funds particular types of projects. These include RECHAR for designated coalfield closure areas, and URBAN for designated deprived urban areas. The sums allocated to the UK through the Structural Funds are substantial, averaging £1.8 billion per year between 1994 and 1999, and are used to match fund resources obtained from UK regeneration initiatives.

The European Commission has stated that approximately £10,000 million of Structural Funds will be allocated to Britain between 1994 and 1999. Much of the funding will flow to urban areas, especially in the older industrial regions. The EU has also made it easier for the private sector to seek support for infrastructural schemes, with limited 'additionality' funds from local authorities.

Box 4.6　Birmingham and EU funding

There have been some European success stories on a local level. In the early 1980s the City of Birmingham lobbied hard to be declared an assisted area in order to qualify for the European Regional Development Fund. In 1984 the city was designated and has subsequently taken a proactive stance towards Europe which has been rewarded with sizeable funding. By 1987 Birmingham had received £78 million from the ERDF and between 1988 and 1991 it gained a £203 million package, consisting of £128 million ERDF money, £31 million worth of European Social Fund monies and £44 million in loans.

It was the early start by Birmingham and its subsequent commitment to Europe which enabled it to gain this funding. Many of the local authorities now active on a European level suffer from late entry into the 'policy networks' of EU funding. However, this problem may be partly solved by the recent introduction of the Single Programme Document (SPD) which should make the process of application for monies simpler, more efficient and enable greater co-ordination.

Securing Resources: Capital and Maintenance

The majority of physical regeneration projects require an element of public sector support. Increasingly such support, be it from the SRB Challenge Fund, English Partnerships Investment Fund or the European Structural Funds (ERDF), will need to be justified in terms of 'hard' outputs – notably jobs saved and/or created (for example, see English Partnerships, 1994). The process of resource allocation from public bodies is becoming increasingly sophisticated as the onus is put on bidders to demonstrate outputs using techniques such as detailed cost–benefit analysis. The pressure to maximise private sector contributions further compounds this and an increasing emphasis is being placed on ex ante project appraisal.

The requirement for match funding from the majority of grant providers (including the National Lottery) also contributes towards increasingly complicated techniques of appraisal as schemes seek to pull together a package of funding from various public and private sources.

As most grants for physical regeneration are for capital expenditure (that is, those awarded by the Millennium Commission) there is a need to prove that once complete new buildings/attractions will be self sustaining in terms of revenue generated for maintenance and running costs. Funding agencies will therefore wish to see evidence of capability to deliver. This is often provided in the form of business plans, details of management arrangements and importantly an early clarification of any legal issues including matters such a land ownership, compulsory purchase orders, joint ventures or contaminated land liabilities. If the ownership of a capital project is likely to change after its completion it will be important for the business plan to include an exit strategy detailing how this is to be successfully achieved. Detailed project planning is therefore becoming increasingly important to success in terms of grant allocation as well as ensuring a scheme's long term economic viability. Project promoters will need to ensure that the total stock of developments does not exceed their capability to finance their continued operation/maintenance. This has been a particular concern of the large capital projects part funded by the Millennium Commission.

One way of overcoming this may be the increased use of private resources through the PFI, which, to date, has been used to secure capital investment in roads, light rail, health facilities and information technology (IT) services among others. Through the formation of private sector–led joint ventures, the PFI transfers financial risk to the private sector. In particular, with the recent extension of the Design Build Finance Operate (DBFO) model to nearly all local authority services, many capital investments using this method of funding will be able to attract favourable revenue support.

Issues to be considered by particular types of project are as follows:

- Landscaping/environmental improvements – usually one-off capital projects requiring some form of ongoing maintenance. Important to allocate resources to maintenance (in form of annualised cost) and also to use materials which will be available in the future should they require replacement.
- Transport projects – will revenue generated from passengers be sufficient to maintain service levels and/or stock? The proposed Birmingham Northern Relief Road aims to cover capital, revenue and maintenance costs from charging users of the road – this is likely to be an increasing trend in the future.
- Attractions (arts/cultural/tourism) – visitor numbers will be crucial as these should create enough revenue to offset running costs.
- Housing – capital improvements to housing stock should by their nature reduce maintenance costs through improved construction and energy efficiency. Transferring redeveloped and refurbished stock to private or housing association ownership will reduce the maintenance burden on public resources.
- Social and economic regeneration (that is, training and business support facilities) – will often depend on continued public sector support in addition to revenue generated to remain viable. Need to be wary of long-term revenue difficulties.

Economic Impact of Regeneration

A range of problems are encountered in attempting to evaluate the economic impact of urban regeneration. It is difficult to examine the economic impact of a specific initiative or programme due to the fact that the outcomes are often the result of the combined effects of several programmes. There is also the counter–factual problem to consider. This is the problem of not being able to predict what would have happened over a similar period of time if no programme was introduced. Despite these problems, evaluations have been able to demonstrate that several projects have proved to be successful. Research conducted into the Urban Development Grant argued that it raised private sector investment in the cities (DoE, 1988b). A similar investigation into the Urban Programme demonstrated that it helped to make environmental improvements and create employment projects (DoE, 1986), whilst evaluation of the Enterprise Zone initiative showed that they enhanced local property markets, removed dereliction, encouraged new business start-ups and created new jobs (HMSO, 1995).

Assessment of Recent Policy

A recent analysis, *Assessing the Impact of Urban Policy* (Robson *et al.*, 1994) by Brian Robson and colleagues as part of the Department of the Environment's Inner Cities Research Programme, provides a detailed examination of the impact of urban policy during the 1980s and early 1990s. As it is the only source to provide such detailed assessment the remainder of this section is, for the most part, reliant on its methods and results.

The impact of urban economic policy in this case was evaluated both quantitatively and qualitatively. The quantitative, or statistical, analysis examined the relationship between the expenditure and the socio-economic outcomes in a sample of 123 English local authorities. Of these 123 authorities, 57 were Urban Priority Areas (UPAs) – the targets of the Action for Cities programme – 40 were similar marginal authorities and the remaining 26 were 'comparator' (i.e. for comparison) authorities.

There were a large number of outcomes that could have been chosen for analysis, but, in this case, the factors considered to demonstrate impact most effectively were the general and long-term rates of unemployment during the period 1983 to 1991, net job changes between 1981 and 1989, the percentage change in small businesses between 1979 and 1990, house price change from 1983 to 1990 and the net change in the number and proportion of 25–34-year-olds between 1981 and 1990. These factors were chosen for analysis as they were easy to interpret and readily accessible and, between them, addressed the areas of employment creation and enhancing the attractiveness of urban areas.

The qualitative information was gathered from two different areas. The first of these was from a survey of the recipients of policy – both residents and employers in inner city areas – and the second came from the outcomes of discussions with experts at the sharp end of policy implementation.

Quantitative Evaluation

A national assessment of 123 local authorities was used by Robson to evaluate impacts and outcomes. With regard to inputs it appears that expenditure in the UPAs shows a lack of fit between the classification of authorities as UPAs and the amounts of resource they received per capita. The UPAs did not receive substantially more per capita than the marginal or comparator areas. Indeed, the poor-performing UPAs, with regard to outcomes, showed large percentage reductions in per capita funding. In considering outcomes, the figures show that overall there were deteriorating conditions in the UPAs over the period as a whole, although some of the individual indicators were positive. Both the general and long-term employment rates show signs of improvement. There were also some successful outcomes in terms of improving residential attractiveness in inner city areas and so some UPAs have been successful in retaining the important 25–34-year-old age group. Urban policy has had the effect, in some UPAs, of slowing down or reversing the erosion of the economic and residential base of some towns and cities. Certain relationships are clear from the above. The main one being that an increase in expenditure in a defined area can be linked to some reduction in overall and long-term unemployment, and also assists in the retention of the 25–34-year-old age group. There is also a link between the increase in expenditure and a positive house price change during 1986–89.

Profiles of UPAs based on input: outcome data reveals that 18 of the 57 UPAs show what might be defined as positive outcomes. Nine of these 18 had high or relatively high inputs, two had a mixed level of inputs and seven had relatively low inputs. A further 18 UPAs showed mixed outcomes and of these eight had mixed inputs and ten had relatively low inputs. Twenty-one UPAs had poor or relatively poor outcomes despite the fact that nine had high and two had mixed inputs. The other ten had low inputs.

Qualitative Evaluation

Turning to the qualitative surveys, 1,299 interviews were conducted with residents of UPAs. The areas of the surveys were selected to enable paired comparisons between areas broadly similar in socio-economic conditions but which had received different levels of financial resources. The main points to come from the surveys were that in areas receiving more money the attitudes of the residents towards the current and future desirability of their area were positive; a form of 'area loyalty' exists, in such areas. This would suggest the need for a more focused social dimension to urban policy to take advantage of this loyalty and sense of community. The respondents outlined their most important issues of concern of which the top three were crime, health care and the cost of living. However, it must be noted that there was a marked difference in responses between racial groups.

Employers' views about urban policy were gleaned by a series of loosely structured discussions. The responses showed that their knowledge of the programmes was limited, even for those companies who appeared to be deriving benefit from schemes. Managers were critical of the multiplicity and bureaucratic impediments they were faced with. Capital-based schemes such as the Derelict Land Grant and the Urban Development Corporations were generally welcomed although access to information was seen to be difficult. Tackling crime in inner city areas was seen as important because extra expenditure on security was a disincentive to locating in an inner urban location.

During interviews and discussions with experts there were three issues continually raised. The first concerned partnerships. The success of partnerships and the programmes they were associated with was determined to a large degree by local consensus. Despite the government aims to encourage and develop these partnerships many respondents felt that much of policy in the 1980s actually undermined the development of partnerships at the local and national level, particularly because of financial restrictions which had reduced the capacity of local authorities to be successful partners. However, the recent City Challenge programme has signalled a change in this trend and the local authorities have started to play an increasing role in partnerships. Respondents felt that the City Challenge programme scheme was innovative and early evidence pointed towards a successful initiative.

The second issue concerns co-ordination and concludes that co-ordination between local authorities, between central government departments and between central and local government is insufficient. Different departments seemed to be operating on different assumptions about the nature of, and, presumably, how to tackle, their problems.

The third issue is policy targeting. There are two types to be considered: targeting resources towards the 57 UPAs, and targeting groups or areas within the UPAs. The responses suggest that there is a need to focus more on groups than on places. This opinion contradicts some of the outcome results which suggest many of the more successful programmes were spatially targeted. There was also considerable support for focusing resources on narrowly defined areas.

The Implications of the Assessment

In many ways a mixed pattern emerges from this assessment. There is certainly evidence to suggest an improvement in terms of unemployment and social deprivation. However, in the most deprived areas it seems that policy has had little effect, indeed, evidence exists which suggests some areas are now worse off. The best situations can be found in smaller, peripheral areas where expenditure has had a positive effect, but this is countered by the fact that there is an increasing polarisation between the 'best' and 'worst' areas.

The aim of creating partnerships has only been partly achieved while the desire to 'lever-in' private investment has had the effect of alienating local communities and authorities in some areas. Local authorities have encountered problems in supporting their role in partnerships because their resources were over-extended and they were unfamiliar with the workings of the private sector. Additionally, in the view of some respondents, private investment has a short-term focus which does little to ensure the stability of urban economic policy or a process of continuous policy development.

The report concludes that much of urban policy in the 1980s lacked coherence and should have been more strategic. Many of the programmes had to be co-ordinated across several governmental departments and this did not work well. The areas where urban policy has worked well are mainly area-based schemes with a community focus and this suggests the need for regional budgets and more local co-ordination. The City Challenge programme reflected this desire as does the SRB.

The study is also critical of the government's economic (property-led regeneration) focus to urban policy, claiming that it has failed to consider important social and community problems and issues, although it does state that some success has been achieved in terms of environmental improvements and physical renewal. However, the report considers that the needs of deprived inner-urban area residents have been ignored and that a major opportunity has been missed to utilise their skills and build local capacity (Robson *et al.*, 1994).

Conclusions from the Robson Report

Box 4.7   Conclusions from the Robson Report

Five conclusions drawn from the Robson Report are designed to influence future urban economic policy:

- There is a need to create effective coalitions of 'actors' within localities and these will prove more effective with the development of structures to encourage long-term partnerships.
- The local authorities must play a significant part in any coalition.
- Local communities must also be involved in these coalitions.
- The coherence of programmes must be improved both within and across governmental departments. Strategic objectives must guide priorities.
- The development of an urban budget administered at a regional level is needed to better reflect local conditions.

As the study was published in 1994 it is worth bringing the situation up to date by considering whether any of the proposals have been met. In terms of the first two points some progress has been made. Across the country a number of new 'partnerships' have been initiated such as the Coventry and Warwickshire Partnership Ltd.

Box 4.8   Coventry and Warwickshire Partnership Ltd

The Coventry and Warwickshire Partnership Ltd was set up in June 1994, as a company limited by guarantee, to provide a formal mechanism for co-ordination and integration of policy initiatives. The establishment of the partnership follows a long history of informal arrangements across a number of local authorities and other organisations designed to support and promote the economic development of the area.

Key partners include the Coventry and Warwickshire local authorities, the Coventry and Warwickshire TEC and the Chambers of Commerce. The partnership has achieved early success particularly in respect of the first round of SRB. A single economic strategy for the subregion has been developed to provide the framework for future economic developed initiatives in the City of Coventry and the surrounding County of Warwickshire.

There has also been a number of developments regarding the organisation and administration of urban economic policy. In April 1994 Government Offices for the Regions were established. These ten integrated regional offices drew together resources from the Departments of Employment, Environment, Transport and Industry and each is accountable to a senior regional director. The intention of the GORs was, according to John Gummer who announced the launch, to initiate 'sweeping measures to shift power from Whitehall to local communities and make the government more responsible to local priorities'. To aid the work of the GORs the SRB was developed which has been previously discussed.

The Future of Economic Regeneration

Over the past three decades economic regeneration has responded to and has been influenced by a wide range of economic, political and social factors. Policy initiatives developed out of a strong economic rationale which in the 1960s and 1970s saw urban areas facing the prospect of continuous decline in the absence of major injections of public expenditure. Many initiatives were designed to overcome the disadvantage of inner city locations in terms of accessibility, environmental quality and the relative cost of land development compared with greenfield sites. In more recent years public sector investment has continued to support economic regeneration with increased emphasis placed on frameworks designed to ensure a more integrated approach emphasising partnership and value for money. The latter reassessment is promoted through a greater reliance upon competitive bidding which seeks to establish clear performance criteria for evaluating outcomes.

This evolving urban economic policy framework seen through the implementation of the SRB programme and the work of English Partnerships places increasing emphasis upon competition, selectivity, community involvement and the implementation of prestigious projects capable of generating a high local impact, both economically and politically. It is also evident that there has been a shift away from traditional industrial and commercial projects to those which have a stronger emphasis on addressing the needs of the local community. For example, many recent projects have incorporated leisure and recreation facilities to meet these needs and as part of wider tourism development strategies have included facilities to attract both day and overnight visitors. As this approach is further developed successful bids for economic regeneration funding will need to:

- involve a broad range of partners;
- address the needs of local communities;
- reflect the sectoral changes occurring in the economy;
- demonstrate innovation;
- point to tangible benefits;
- be capable of replication.

There is evidence, through recent funding decisions, that flagship projects are beginning to emerge as the preferred approach to address the major challenges facing towns and cities in the twenty-first century. Proposals which address issues such as the wider use of technology and telecommunications, transport and communications, economic and environmental sustainability and promote urban living, for example through

---

Key Issues and Actions

- Recently, due to changes in the workings of urban and regional economies and the increasing globalisation of markets and economic and industrial restructuring, cities have been in decline.
- Economic regeneration is a vital process in urban regeneration.
- Urban regeneration aims to attract and stimulate investment, create employment opportunities and improve the environment of cities.
- Funding for schemes and programmes comes from a wide variety of sources with increasing competition for limited resources.
- Urban economic regeneration involves a partnership created from national and local government, the private and voluntary sectors and members of local communities.
- The developing role of Regional Development Agencies in economic and urban regeneration.
- Urban economic policy must continue to be dynamic and responsive to changing circumstances.
- Examples of best practice in partnership arrangements should be widely publicised.
- Potential fragmentation of urban policy places an increasing role on the development of a clear strategic context and vision of the future.
- It is essential that the case for urban regeneration funding is set within a wider cost/benefit context and its role in sustainable development at the national and international level is fully recognised.

---

mixed-use developments, are likely to be favoured. Urban economic policy is increasingly concerned with the efficient use of financial resources rather than simply promoting the maximisation of economic growth. From a regional economic development perspective the developing role of the newly formed Regional Development Agencies as promoters and co-ordinators of economic and urban regeneration will be an important issue for practitioners over the coming years.

References

Atkinson, R. and Moon, G. (1994) *Urban Policy in Britain*, Public Policy and Politics Series, Macmillan, London.

Audit Commission (1989) *Urban Regeneration and Economic Development: The Local Authority Dimension*, HMSO, London.

Banfield, E. (1970) *The Unheavenly City*, Little, Brown & Co., Boston.

Department of the Environment (DoE) (1986) *Assessment of the Employment Effects of Economic Development Projects Funded by the Urban Programme*, HMSO, London.

Department of the Environment (DoE) (1988a) *City Grant Guidance Notes*, HMSO, London.

Department of the Environment (DoE) (1988b) *Urban Policy and DoE Programmes*, HMSO, London.

Department of the Environment (DoE) (1991) *General Note: Derelict Land Grants*, HMSO, London.

Heseltine, M. (1983) *Reviving the Inner Cities*, Conservative Political Centre, London.

HMSO (1977) *Policy for the Inner Cities*, Cmnd 6845, HMSO, London.

HMSO (1987) *An evaluation of the Enterprise Zone Experiment*, HMSO, London.

HMSO (1988) *Action for Cities*, HMSO, London.

HMSO (1995) *Final Evaluation of Enterprise Zones*, HMSO, London.

Imrie, R. and Thomas, H. (eds) (1993) *British Urban Policy and the Urban Development Corporations*, Paul Chapman, London.

Lawless, P. (1988) British inner urban policy: a review, *Regional Studies*, Vol. 22, no. 6, pp. 531–42.

National Audit Office (1990) *Regenerating the Inner Cities*, HMSO, London.

Oatley, N. (1995a) Competitive urban policy and the regeneration game, *Town Planning Review*, Vol. 66, no. 1, pp. 1–14.

Oatley, N. (1995b) Urban regeneration, *Planning Practice and Research*, Vol. 10, no. 3–4, pp. 261–70.

Public Accounts Committee (1989) *Twentieth Report: Urban Development Corporations*, HMSO, London.

Richardson, H. (1971) *Urban Economics*, Penguin, Middlesex.

Robson, B. (1989) *Those Inner Cities*, Clarendon, Oxford.

Robson, B., Bradford, M., Deas, I., Hall, E., Harrison, E., Parkinson, M., Evans, R., Garside, P. and Robinson, F. (1994) (the Robson Report), *Assessing the Impact of Urban Policy*, HMSO, London.

5 Physical and Environmental Aspects

Paul Jeffrey and John Pounder

Introduction

The physical appearance and environmental quality of cities and neighbourhoods are highly potent symbols of their prosperity and of the quality of life and confidence of their enterprises and citizens. Run-down housing estates, tracts of vacant land and derelict factories, and decaying city centres are the all too visible faces of poverty and economic decline. More often than not they are the symptoms of decline or of a town's inability to adapt quickly enough to rapid social and economic change. However, inefficient and inappropriate infrastructure or worn-out and obsolescent buildings can be a cause of decline in their own right. They fail to serve the needs of enterprises in new and growing sectors and impose costs in use and repair which are higher than average and beyond the means of those in poverty or firms on the margins of profitability. They blight the investments, property values and confidence of those living or working nearby.

Likewise, environmental decay and a neglect of the fundamental principles of sound resource use can damage both the functioning and reputation of a city. Above and beyond this, the ecological 'footprint' or 'shadow' of an urban area frequently extends beyond the administrative boundaries of a city and reflects the consumption of resources associated with urban living.

Physical renewal is usually a necessary if not sufficient condition for successful regeneration. In some circumstances it may be the main engine of regeneration. In almost all cases it is an important visible sign of commitment to change and improvement. The key to successful physical regeneration is to understand the constraints and the potential of the existing physical stock and the role improvement can play in enabling, and where appropriate promoting, renewal at regional, urban or neighbourhood level. Successful realisation of the potential requires an implementation strategy which recognises and takes advantage of the changes under way in economic and social activity, funding regimes, ownership, institutional arrangements, policy and emerging visions of urban life, and the roles of cities.

This chapter addresses these key themes in physical and environmental regeneration:

- the components of the physical stock;
- socio-economic changes and the new requirements on the physical stock;
- the roles of physical regeneration in urban regeneration;
- assessing the stock;
- developing solutions and schemes;
- funding, capital and maintenance;
- involving the actors, participation, defining delivery mechanisms and institutional arrangements;
- the changing context of funding, institutions, and policy;
- new visions of urban areas.

The Physical Stock and Socio-Economic Change

The Components of the Physical Stock

In urban regeneration it is very often the state of the buildings which dominates an appreciation of the physical conditions of the area to be addressed. However, the components of the physical stock go much wider than this and whilst action may not be necessary across all components of the stock there is at the least a need to assess all aspects of the stock. This includes:

- buildings;
- land and sites;
- urban spaces;
- open spaces and water;
- utilities and services;
- telecommunications;
- transport infrastructure;
- environmental quality.

Components such as the available utilities and services and the telecommunications and transport infrastructure can be critical elements in regeneration. For example in some of the older industrial parts of the UK, increasing the demand for gas supplies will require that the existing gas mains delivery network is strengthened; the London Docklands' Development Corporation (LDDC) found that it had to take the initiative in investing in electricity substations in some areas to provide new capacity before new developments could be encouraged.

The quality of telecommunications available is becoming increasingly important for firms and hence the availability of Integrated Services Digitial Network (ISDN) lines, compatible exchanges and, to a lesser extent, fibre optics links are all relevant considerations for physical renewal. Similarly, accessibility is coming to have a wider scope. The three traditional qualities of a site – location, location, location – emphasise this, but accessibility is coming to be more than easy connections to a high-quality road network – important as that is for certain sectors. Planning Policy Guidance (PPG) 13, and indeed the preoccupation of many regeneration schemes in inner city or town centre areas, means that there is a need to ensure that access by mass transport systems is viable as the pressures of congestion and traffic pollution result in restrictions on car use. Furthermore, some of the growing sectors which might be relied upon to help regeneration (such as tradable business services, software houses, specialised and high value-added manufacturing) operate in national and international markets and require high-quality rail and air links to their market places and other sites. Thus the quality of rail services and the suitability of airport capacity may be relevant considerations, depending upon the spatial scale of the area being addressed for regeneration and the possible future economic structure envisioned in the programme.

The impact of transport infrastructure can have a negative as much as a positive impact as transport requirements change. For example, the City of Birmingham has recognised that its inner ring road is damaging the quality of the centre of the city and inhibiting visitors and the expansion of the centre. Hence, as part of its regeneration scheme which includes the Symphony Hall and Centennial Square it has lowered the road below the level of pedestrian movements; and in some of its latest schemes is considering removing the road altogether. In the case of the proposals for the Masshouse Circus and new Millennium Point regeneration scheme this entails the highly costly process of demolishing a prestressed concrete elevated road as well as finding a way of dealing with all of the displaced traffic.

Environmental quality is now recognised as a key and necessary component in the locational decisions of many firms and high-skilled workers. It is therefore important from an economic development point of view as much as from the perspective of giving residents a good quality of life and expressing confidence in an area. Environmental quality is an integral feature of the buildings – new and old – and of the urban and natural spaces. However, it is important to address it in its own right. Whilst the overall form and structure of an area may offer significant possibilities – the existence of canals and water is the now 'old hat' example – the reverse may also be the case. For example, some now obsolete mining villages in the East Durham Coalfields and elsewhere, where the overall structure of the village is of poor quality buildings and a drab layout, can be dramatically altered by major planting and possibly selective demolition. In more detail, use of open water as a focus for regeneration may require extensive work to clean up and maintain water quality – as in Salford Quays – or air pollution from poorly controlled factories may be an inhibition to new development as was the case in certain parts of the Black Country which were affected by foundry emissions. Such problems are by no means cheap to resolve, yet they can be critical to the success of a scheme and therefore need to be taken into account early in the development of the regeneration strategy.

The criteria used for an initial assessment of the stock will depend not only upon current standards and obvious measures, such as dereliction, but also to some degree upon an appreciation of the future uses of and vision for the area. To that extent the assessment has to be an integral part of the process of developing a solution for the area and be based upon an understanding of the way in which changing economic and social demands will require different standards in the future. Furthermore, the scope and detail of the attention to the stock will vary depending upon the spatial scale of the regeneration project – from single site to whole urban area. When considering the role and state of the physical stock in the regeneration of an area as large as East Manchester or the London Docklands, then it is important to include all of the elements outlined above and, necessarily, the assessment of sites and buildings has to be done, at least initially, using general criteria and features which can be readily assessed in a simple survey. On the other hand, when dealing with the regeneration of a smaller area composed of a few sites, then the focus is likely to be upon the sites and buildings with the quality and capacity of the other aspects of the stock being a constraint or limiting factor rather than being something to be changed or addressed in the strategy.

Socio-economic Change and the Stock

The rate of socio-economic change is much more rapid than that of the physical stock. Indeed it is sometimes this which causes the problems of decline and certainly adds to the costs of regeneration. It is fundamental that any regeneration scheme is built upon realistic and sustainable social and economic trends, and therefore that the consideration of the stock – its suitability, strengths, weaknesses and what is required in the future – is undertaken within the context of well-founded understanding of what the market, economic and social conditions will be like. For example it is not possible to appraise the office stock if there is no clear understanding of the likely – or desired – types of commercial activity that will be encouraged to develop in the area. The requirements of 'back office functions' for large business services are quite different from the offices suitable for legal and professional practices.

Therefore in the early stages of regeneration work there needs to be close integration between the developing scenarios for the role and economic and social functioning of the area and the physical appraisal and proposals. The fact that we are at a period when there might be some quite radical changes in both socio-economic preferences and in the way we use and develop the physical stock makes the exercise of forward investment more challenging and exciting. We would appear to be at a point where there might well be a major shift in attitudes to the use of the car in cities and, therefore, to the development and funding of public transport systems. Some cities are seeing a renewed interest in city-centre living and the changes in the demographic structure will alter the demand for different types of housing and facilities. The shift in the mode of delivery to a more community-oriented, partnership approach is well established, but the development of agencies such as English Partnerships and the remit and powers they have will significantly affect the types of regeneration that can get funded. Beyond this, the longer-term impact of telecommunications on financial services and their location (possibly reducing the need for retail outlets), on home-working patterns and on shopping and leisure may well be dramatic for urban areas.

SWOT Analyses

Appraising the Stock

At an early stage during the process of appraising an area and developing ideas for a regeneration strategy it is helpful to provide a quick appraisal of the physical stock. This might address very generally whether it currently acts as a constraint on the development of the indigenous enterprises and the households in the area, and whether there are any obvious major gaps in capacity of the quantity and quality that are generally thought to be necessary to support modern forms of development. Clearly, in certain cases, such as schemes focused on poor housing, many of these constraints and problems will be clear. It is also useful to have a rapid appreciation of the strong features and potentialities in the area upon which initiatives might be built (natural features, waterscapes, historic or architecturally interesting buildings, etc.). Such a rapid appreciation can then be put alongside the emerging understanding of the economic strengths and potential roles for the area and can help in the development of an overall strategy.

The Roles of Physical Regeneration in Urban Regeneration

Physical regeneration, like all forms of economic development, has been subject to fashions over time – for example, the fashion for 'housing leading to regeneration' or 'flagship projects'. Thus to some extent there has been a change of emphasis from one type of 'solution' to another. This is rather less a progression to ever more suitable approaches than a succession of different approaches – each of which has its role depending upon the problems of the area to be regenerated, the influence of the policy context in force at the time and its impact on funding, and the current market conditions. It is therefore useful to look at each of the potential roles separately as they may each be appropriate in the right circumstances.

There are at least five different roles which physical regeneration may play in the full regeneration of an area:

Box 5.1   Liverpool Central Business District Action Plan

The first stage in developing a co-ordinated and managed programme of development, transportation, environment and marketing proposals for Liverpool's Central Business District (CBD) was an analysis of the existing stock. The appraisal was set in the context of current and emerging demand and supply trends in the national and regional office market, and set out to determine how the existing physical stock helped or hindered the perceived role of the CBD. The appraisal comprised assessments of:

- the 'office product' currently being offered:
  – available space
  – amount of quality space
  – flexibility/functionality of space
  – age of buildings
  – occupancy levels;
- urban design and urban environment:
  – building design
  – character of conservation area(s) and listed buildings
  – function of the area
  – identification of weaknesses;
- transport infrastructure and services
  – road network and external access
  – traffic flows
  – public transport capacity
  – parking provision.

- removing constraints;
- leading the change;
- building on opportunities;
- supply side investments;
- integrated socio-economic and physical renewal.

Which of these roles is the most important in any given scheme is to some extent dependent on the future vision for the area – the 'solution' – and in part upon the underlying problems which gave rise to the decline and need for regeneration in the first place.

However, they can be related to the underlying strategy assumptions upon which the particular regeneration strategy is based. There are a limited number of these basic approaches which relate to where the drive for regeneration and the ongoing investment will come from. These are:

- Build upon and develop the potential of the indigenous firms and skills in the area: in this case particular attention needs to be paid to the constraints that the physical stock is currently placing on local firms and any major lacks in the provision of infrastructure and services etc.
- Unlock latent demand and expenditure from current users and visitors to the area: in this case new forms of activity, supply-side actions such as the provision of museums etc. are relevant.

- Attract inward investment; in which case there is the need to ensure that the quality of the environment and the facilities and the quality of infrastructure, services, sites and buildings is adequate to compete with those offered elsewhere.
- Attract new visitors to the area: in which case the quality of the environment, the development of visitor attractors – which may need to be supply-side led – and the refurbishment of existing features and buildings of interest are the key approaches.
- Build upon the strengths of the members of the community: in which case the focus is likely to be on those types of activity such as housing renovation, estate development which are an integral part of an integrated local development strategy.

Removing Constraints

A classic and common constraint has been derelict and contaminated sites where the costs of bringing the site to a state where it can be redeveloped is very high yet market demand is weak. Such problems have been at the heart of the activities of many of the Urban Development Corporations. In the Black Country Urban Development Area, the Black Country Development Corporation (BCDC) considered the condition of each potential development site and costs of reclamation as key determinants in the allocation of proposed uses. Reclamation could then be tailored to the most appropriate land use for a particular site. In the absence of any statutory standards to which land should be reclaimed the BCDC worked with the National Rivers Authority (NRA – now subsumed within the Environment Agency) to agree standards for seepage of contaminants for sites being reclaimed. This gave the NRA confidence in the reclamation activities of the corporation and gave the BCDC and its developers advance knowledge of the extent of works required and associated costs before the works began therefore reducing uncertainty. The choice and standard of remediation can be fundamental to financial viability and acceptability in planning terms. Clarifying these issues at an early stage (i.e. as part of the planning permission or through a Section 106 agreement, rather than as a condition on a permission to be determined after approval) is crucial.

Other less obvious constraints to regeneration occur where the land is held in many small sites, unsuitable for current developments and yet difficult and costly to assemble – very often because the structure of industries, and therefore industrial and commercial sites and properties, were based around a large number of small firms. These may still be in existence but finding their current premises inefficient for modern practices (e.g. multi-storey factories, or sites on narrow roads with little room for parking, heavy goods vehicle (HGV) delivery, or on sites with no room for *in situ* expansion). As such areas decline, successively less profitable uses are made of the locations – scrap metal and car breakers, car repair, HGV storage, etc. come to operate out of such areas. They pose two problems. They further reduce the book value of the property of the existing firms – thus making it difficult for them to borrow and invest (and encouraging them to move away). Second, they represent a major potential cost to an agency or local authority that wishes to encourage redevelopment of the area because the costs of relocation are significant. Such problems are classically found in the older industrial areas of the UK and underpin a large part of the strategies of the Urban Development Corporations, and still underpin the operation of English Partnerships when it helps development through supporting funding of derelict land reclamation.

As well as building and site configuration there may be other constraints such as poor road layout and access, again a common feature of inner city industrial and commercial quarters. There are many areas in which sites are inaccessible and therefore a key response is to open up areas with new spine roads such as those in several of the UDCs, including the St Phillips Causeway (Bristol) and the Black Country New Road (Black Country – see Box 5.2).

Box 5.2   Black Country New Road

The importance attached to good transport infrastructure in terms of attracting inward investment is no more clearly exemplified than in the case of the construction of the Black Country New Road (BCNR). The completion of the road, which runs through the heart of the Urban Development Area (UDA) and provides a direct link between Junction 1 of the M5 with Junction 10 of the M6, was seen as a crucial prerequisite for the economic regeneration of this part of the West Midlands. Access to the strategic motorway network is crucial to modern industry and the corporation realised that quick access to the network would be vital for the attraction of modern industries employing Just In Time principles. Indeed the BCNR formed a central ingredient in the corporation's marketing strategy and promotional material to potential investors/ occupiers.

In Birmingham a different type of constraint has been met – that is the inner ring road which stands as an elevated prestressed structure between the city centre and one of the key inner city areas requiring regeneration. The inner ring constrains the city-centre functions from easily expanding outwards in this area. In Belfast the constraints are slightly different in that there is little available land on which to encourage the industrial development associated with the docks and, therefore, sites have had to be reclaimed from the estuary.

At a larger spatial scale, there may be constraints in terms of the capacity of the regional road system's ability to grow in response to forecast growth. Such constraints can be relieved through the introduction of mass transit systems. Overcoming the potential constraint of future congestion on further development of the region was one of the reasons for public sector support of the proposed West Midlands Metro.

Leading the Change

The very visible nature of new developments means that they can play a significant role in establishing a changed image for an area, act as a declaration of intent and vote of confidence in the area and so bring in other developments behind them. Such so-called 'flagship projects' were at one stage highly fashionable. A classic example of this is the Birmingham Symphony Hall and Convention Centre which established the clear intent to support and encourage regeneration of a whole swathe of derelict and underused land to the west of the inner city. A key feature of such approaches is that they are large enough to have a major visible impact on the immediate area and so change the context for further investment by reducing the negative image and run-down nature of the area. In Birmingham this physical transformation of part of the city centre through the development of key prestige projects was fundamental to the reconstruction of the image of the city, both nationally and internationally, as a centre for business tourism (Loftman and Nevin, 1996).

Flagship projects were characteristic of the approach to urban regeneration taken by many of the Urban Development Corporations set up in the 1980s. Well-known examples include Canary Wharf (London Docklands' Development Corporation) and Albert Dock (Merseyside Development Corporation). The aim of these schemes was to boost confidence in the designated Urban Development Area and thereby help stimulate private sector investment.

Physical development can lead regeneration in less dramatic ways. For example it may be that the most effective way of beginning to mobilise community development in poor housing estates is to begin by improving the physical condition of the housing itself. Examples of this are: Vauxhall Urban Village, Liverpool; The Eldonians, Liverpool; Hulme, Manchester; and Miles Platting, Manchester. One of the interesting aspects of this approach is that it gives the opportunity to involve the residents directly in the development plans and, through community development schemes, to involve them in the work through training and work experience schemes.

Box 5.3   Vauxhall Urban Village, Liverpool

A partnership between the resident communities and the Merseyside Development Corporation. Began as a plan to build new housing and developed through a 'bottom-up' approach to regeneration into a comprehensive project to provide community facilities, training opportunities and local enterprise initiatives.

Box 5.4   Miles Platting, Manchester

The refurbishment of a high-density public housing estate (originally consisting of 4,350 units). By early 1980s the estate was characterised by the typical environmental and social problems associated with 1950/1960s housing estates, and heightened by high rates of unemployment. It was recognised at the outset that significant physical refurbishment without complementary environmental improvements would have a limited impact on the low levels of confidence the residents had in the estate. Extensive community involvement was crucial to the successful development of the project. Involvement was promoted on the estate as a whole and within individual housing blocks, allowing residents to discuss the details of policy. A result has been reduced vandalism, a greater level of commitment and involvement by residents and a lessening of the hostility felt by them towards the estate.

(Source: ECOTEC, 1988)

Box 5.5   Manchester – Hulme homes for Hulme people

Hulme covers an area of 272 acres (equal to that of Manchester city centre). The regeneration strategy focuses on a housing programme planned and implemented by a partnership comprising a group of housing associations, local residents, Hulme Regeneration Ltd, Manchester City Council and private sector developers. The redevelopment has been led by a successful City Challenge bid and over a period of five years more than £150 million of public and private funds have been invested in the demolition of unfit dwellings and the building and renovation of almost 2,000 homes. The housing programme has laid the basis for a broader regeneration initiative (drawing on SRB, Capital Challenge and URBAN initiative funds), including community development schemes, by dramatically altering perceptions of the area.

Building on Opportunities

In some cases the physical stock itself has strong qualities which can be developed and built upon. The most well-known example of this is the use of water, docks, canals etc. as a feature around which to establish and encourage development. Since the ideas were successfully developed in Baltimore some 15 years ago there are now numerous examples of this approach world-wide. In the UK, there are many examples of housing developments featuring canalside locations and indeed British Waterways has successfully established a large number of regeneration projects building on the asset of their canalside sites. An example is the Coalisland (Northern Ireland) project which won a BURA best practice award in 1996. It represents the harnessing of industrial heritage to create a focus for economic and social growth; the result being the creation of 50 workspace units, the formation of six new industrial sites and extensive environmental improvements. The conversion of a redundant cornmill into a community

---

Box 5.6   Salford Quays

A local authority-led physical redevelopment aimed at creating a unique character based on a the creation of a new quarter and its relationship to water. Prior to development the land had to be cleared of outdated/obsolete buildings and infrastructure and a new canal, road and service structure installed. The realisation that the Greater Manchester light rail scheme could be extended to serve the quays enhanced recognition of the opportunity that existed for large-scale commercial arts, sports and leisure facilities, which are currently being developed.

---

centre is a catalyst for further development. Real community involvement has led to a feeling of ownership which should ensure its sustainability.

There are of course a much wider range of opportunities potentially offered by the physical stock in addition to the water feature which has become almost a cliché. These include buildings of architectural or historic interest. Well before the Central Manchester Development corporation was established, Manchester City Council had initiated the Castlefield project in a derelict area between the city centre and the Rochdale canal. Here there are the archaeological remains of a Roman fort which have been used as the basis of an urban park and local improvement which has encouraged other investment to come into the area, such as a hotel and new, small office developments.

Supply-Side Investments

Supply-side investments range from the traditional serviced site and building provision, transport access through to specialist investment such as cold stores on transportation parks (such as the example at Hull), or tourist destinations (such as museums and hotels). In many instances the so-called flagship projects have sought to stimulate economic activity and attractiveness of an area by supplying services or tourism and visitor destinations which were not available and hence for which demand was suppressed. The Albert Dock on the Liverpool waterfront is a good example of a flagship regeneration scheme which combines a waterside location with the refurbishment of a listed building, and has also created a market for tourism which did not previously exist, attracting up to 6 million visitors per annum.

The LDDC undertook significant preinstallation of services and telecomms capacity with respect to electricity substations and satellite links. They even went so far as to fund the capital expenditure for building new schools in areas they were promoting for housing development so that the facilities were available as an attraction to new in-movers. The LDDC has, however, also been criticised for not providing basic transport infrastructure in advance of its initial development projects. This was exemplified by

---

Box 5.7   Sutton Harbour Regeneration Scheme, Plymouth

Winner of the Secretary of State for Environment Award for Partnership in Regeneration 1996.

During the 1980s Sutton Harbour had fallen into serious decline. The key elements of a regeneration strategy were the construction of an entrance lock to overcome tidal restrictions, the development of a modern fisheries complex, major environmental improvements and the creation of new links with the historical core of the city. The physical and operational improvements to the harbour have: opened up previously derelict land to investment; encouraged the diversification of industries; enhanced the commercial viability of a major investment in the National Maritime Aquarium; and, through new links with the historic city core, further boosted the local economy through tourism.

---

the phasing of the Limehouse Link road serving Canary Wharf and the delays to the completion of the Jubilee Line extension.

Integrated Socio-Economic and Physical Renewal

After the concentration on physical, property-led regeneration initiatives during the 1980s, exemplified by the UDCs, the advent of initiatives such as City Challenge represented a new attempt at bringing together the property and socio-economic aspects of urban regeneration. In Leicester for example, the local community (under the umbrella of the Bede Island Community Association – BICA) have representation on the Challenge Board. This process of involvement has been further extended with the advent of the Single Regeneration Budget Challenge Fund which requires bids for funding to be submitted by partnerships of private, public and community sectors, thereby integrating a broader range of interests than may have previously been the case with statutorily imposed regeneration bodes such as the UDCs. Examples of initiatives which combine both aspects of renewal are the Shirebrook and District Development Trust, and the regeneration of the Roundshaw Estate in South London. The latter exemplifying how the advent of new funding mechanisms (in this case the SRB Challenge Fund) has brought together a wide variety of interest groups and stakeholders for the purposes of developing and implementing regeneration.

The European Commission's Urban Pilot Projects (UPP) are one of a range of Community Initiatives funded from the EU's Structural Funds and are designed to explore and illustrate innovative approaches to tackling urban problems of environmental and industrial decay and social exclusion. The UPP in Bilboa, Spain is a good example of integration of physical and social regeneration. Focused on the highly disadvantaged and physically, socially and economically isolated district of La Vieja, the UPP actions concentrate on the rehabilitation of the area's infrastructure

---

Box 5.8   Shirebrook and District Development Trust

Operates in the heart of the former Derbyshire Coalfield, providing advice to local people; supporting local business; promoting the town centre and renovating prominent derelict properties. The executive comprises local volunteers who represent the interests of the local community. It has developed its own assets through property purchase (which provides a long-term income stream) whilst working towards the reintegration of the local workforce into the broader economy.

Box 5.9   The Roundshaw Estate, London Borough of Sutton

The Roundshaw Estate is a 1960s development, physically and visually cut off from the rest of the community, in need of extensive redesign and refurbishment. The regeneration of the estate has been enabled by a successful SRB bid by the Sutton Area Regeneration Partnership. The partnership comprises: residents of the estate; the business community, public sector agencies, and voluntary organisations; the Local Agenda 21 Forum; and health professionals. The private sector takes the lead in the regeneration programme and subgroups are directly responsible for the delivery of each of the strategic objectives, which include:

- helping generate sustainable economic growth in the area;
- improving living conditions, quality of life and environment;
- providing a comprehensive range of education and training;
- fighting crime and fear of crime;
- improving access and transport.

Hence, although the strategy comprises significant housing development and refurbishment it recognises that the physical redesign of the built environment will not in itself bring about the economic regeneration of the area, and so must be accompanied by improved facilities for health care, education and training and transport.

Box 5.10   Wise Group

The Wise Group started in Glasgow in the mid-1980s and aims to improve the physical environment of poorer communities and reduce unemployment by providing improved services to low-income households. The group comprises a series of not for profit companies carrying out physical regeneration activity (including whole house refurbishment) on social housing estates by recruiting and training a workforce drawn from the long-term unemployed. Trainee workers are given a one-year programme of work and after an initial eight-week training period are paid a day rate for the job which takes them out of the benefit system. Core businesses comprise: Heatwise which concentrates on home insulation, efficient heating and home security; and Landwise which concentrates on environmental upgrading.

Box 5.11   Housing Action Trusts – Castle Vale, Birmingham

Six Housing Action Trusts (HATs) have been set up by local authorities since 1991, to take over ownership and management of large council estates with high levels of structural decay and social problems, in order to bring about investment and regeneration which is beyond the scope of local authority budgets. Castle Vale HAT in Birmingham, formed in 1994, is responsible for the regeneration of Castle Vale housing estate. Built in the 1960s Castle Vale comprises 34 high-rise blocks originally planned to house 22,000 people displaced by inner city slum clearance. Over five years all but three of the high-rise blocks are scheduled for demolition and replacement. A key aim of the plan is to encourage residents to consider forms of tenure other than public, with ownership of the new homes to be transferred to various non-profit making housing associations. Indeed this has proved necessary simply for achieving the income required for completing the plan in the absence of adequate funding from central government. Castle Vale and other HATs illustrate the massive physical regeneration required on housing estates throughout the UK and the importance of radical physical transformation in eliminating the stigma attached to many of these areas. However, economic regeneration is more difficult than physical regeneration. In recognition of this it is working with local employers to develop training schemes to meet future skills requirements so local residents will be well placed to take advantage of future employment opportunities.

---

including, for example, the provision of a health and hygiene centre. In addition the project embraces the concept of self-rehabilitation, with most renewal works having been carried out by the local labour force and residents (following practical training), and are specifically targeted at creating opportunities within the area.

Developing Solutions and Avoiding Conflict

Working with the Market

It is fundamental that physical regeneration strategies understand the property requirements of the types of firms they are aiming to attract, due to the increasing differentiation and segmentation of property markets. This is becoming more important with the increasingly rapid development and employment of information and communication technologies and their implications for infrastructure requirements.

Physical regeneration is therefore important to correct market failures, where there is a mismatch between supply and demand for land or property. This is often exemplified by the reluctance of institutions to invest in brownfield sites. This weakness in the operation of land and property markets is recognised by the European Commission which, despite a general move away from financing physical infrastructure investment, still insists on the inclusion of 'Strategic Development Sites' as a priority measure in the Single Programming Documents informing expenditure of Structural Funds for regional development in Objective 2 regions. Without public intervention, the demand for land and property would increasingly be channelled into modern, flexible, accessible greenfield sites which do nothing for the problems associated with inner urban industrial decline and unemployment. Hence, for the purposes of urban regeneration strategies it is important to develop strategic assessments of land and buildings supply and demand characteristics which can identify: the quality of existing stock; the sources of future demand by industrial sector; and property trends in terms of location and type of premises. Steps can then be identified by which public investment can contribute towards correcting the mismatch between demand and supply, which may arise because of: a shortage of a particular type of land; sites not being viable for employment use; or sites requiring upgrading in order to meet investor/occupier expectations. Re-commendations for public investment could include land reclamation, infrastructure provision or property development.

Displacement and the Need for Co-ordination

Turok (1992) highlights the tendency for new property to be taken up by existing local firms relocating from elsewhere in the urban area, taking advantage of available subsidies and thereby displacing vacancies to older units in the urban region. This would suggest a need to formulate city-wide (as opposed to site-specific) approaches to transforming/updating vacant sites if a net overall economic gain is to be achieved.

Focusing policy instruments in concentrated areas (as exemplified by Enterprise Zones and Urban Development Corporations) also gives rise to certain other dangers, identified by Healey (1995). First, there is a danger of clustering activity and consequently distorting land and property markets in favour of certain areas. Second, in the absence of a co-ordinated strategic approach, competition can arise between different agencies which are promoting different parts of the urban region. Healey gives the example of the conflict between Royal Quays in Newcastle with an adjacent City Challenge area. A consequence can be the undermining of central area (brownfield) proposals in favour of more peripheral (greenfield) sites. Healey concludes that the co-ordination of infra-structure provision, land-use planning policy and subsidies for property development is a necessary prerequisite for overcoming these problems associated with physical regeneration. Indeed the proposed Regional Development Agencies may provide an opportunity to improve such co-ordination at a regional level.

Environment and Urban Regeneration

As indicated above, the need for physical regeneration would imply some form of environmental improvement. A run-down urban area is likely to be manifested in physical dereliction, including vacant and derelict buildings and land, and perhaps contamination. These are environmental problems, both aesthetically and physically, and will have major impacts on the perception of an area held by potential investors and the ability of the area to market its assets. However, it is becoming increasingly necessary to justify environmental improvement investment programmes in terms of the economic benefits which will accrue as a result. In other words, for the purposes of securing regeneration resources there needs to be an economic rationale for environmental improvement works. Furthermore, the enhancement of the environment can prove beneficial for both business and the local community (Roberts, 1995).

In recent years numerous urban regeneration initiatives have included extensive elements of environmental improvement, largely for the purposes of attracting private sector investment. Such initiatives include amenity improvements (e.g. landscaping and planting), ground treatment (i.e. land assembly, acquisition, clearance and sale) and improved site access and services. An increasingly important element is the quality of urban design, as has been exemplified in the competitions held for the master planning of Manchester city centre following extensive bomb damage. The nature of environmental improvements will depend on the requirements of individual locations and the end uses they are seeking to encourage (as outlined earlier). Residential, leisure and industrial uses tend to be focused on their immediate surrounds and the particular site in question. Retailing and higher-value commercial uses are more sensitive to their wider environs.

In a study for British Waterways, ECOTEC (1996) attempted to establish the economic benefits of investment in the canal network and develop a mechanism for appraising the benefit as one tool in deciding on an investment programme. Using a case study approach, the research also identified the relationship between canal improvement schemes and the regeneration of adjacent urban and rural areas (see Box 5.12).

The sensitivity of retail and office uses to their surrounding environment has been a driving force behind the appointment of town centre managers and, more recently, the formation of semi-autonomous town centre management companies (for example, Coventry's City Centre Co.) and the promotion of Town Improvement Zones (TIZs). These public–private partnerships have been formed for the purposes of improving town centre environments in response to the threat to central urban areas from out of town developments. There has been a realisation that town centres need to offer a high-quality shopping and working environment in order to remain competitive.

Box 5.12   Gas Street Basin/Brindleyplace, Birmingham

A recent example of comprehensive land assembly and development promotion by Birmingham City Council, incorporating towpath and bridge replacement, canalside engineering works and improvements to listed structures. This pump-priming activity has brought about considerable levels of new investment in mixed-use development incorporating office, retail, leisure and housing sectors. The mixed-use element has been critical to its success with its high-quality canalside setting and its situation in relation to the International Convention Centre and the National Indoor Arena has also been significant. However, the canalside improvements have been criticised for encouraging the spread of economic activity via new pedestrian linkages – in an area characterised by heavily used roads. The insistence on leisure uses occupying units on the canalside has resulted in the canal itself being the focus of activity.

Sustainable Development, Urban Form and Regeneration

In recent years the term 'environment' has come to encompass a much broader range of issues than those indicated above. These 'global' environmental concerns include, the efficient use of resources, biodiversity, air quality, the depletion of the ozone layer and global warming. Urban areas have a large role to play in tackling these global issues, and in doing so moving towards more sustainable forms of development. The notion of the compact city as a sustainable built form is becoming increasingly popular among policy-makers at a national and European level: the implications of this include the promotion of central area revitalisation, higher urban densities, and mixed-use developments – all of which seek to reduce the need for travel. Indeed these principles are now embedded in UK government planning policy (notably PPG13 and PPG6).

Urban Housing

The publication in 1996 of a DoE discussion paper on household and housing growth (UK Government, 1996) added further impetus to the debate about sustainable urban form. The paper included a suggested target of 60 per cent of all new residential development should take place on reused urban (brownfield) land by the year 2005. This figure was set in the context of the government's own projections of household growth which suggest that for the 25-year period 1991–2016 it will comprise 4.4 million households. This generated much public debate about the target figure which should be adopted, with many environmental campaign groups arguing for much higher targets.

The target of 60 per cent (of new homes to be built on previously developed land over the next ten years) has subsequently been confirmed in the White Paper, *Planning for the Communities of the Future* (DETR, 1998), produced by the current government. This proposes that Regional Planning Conferences should be given greater responsibility for identifying targets in terms of extra housing:

- on all previously developed land in urban and rural areas;
- on previously developed land in urban areas.

Breheny (1997) suggests that the increased pressure to reuse brownfield land is likely to have a major impact on urban regeneration policies and practice. However, as he explains, the desire to maximise the proportion of housing developed on brownfield land will prove difficult for technical, economic and political reasons.

- *Economic*: It will be necessary to reverse the trends of counter-urbanisation of population and decentralisation of industry which have taken place in the UK over the last 30 years. This implies a need to destroy the negative perceptions associated with many inner urban areas, and will heighten the need for the creation of positive images and environments in such areas. Perhaps the greatest obstacle to achieving a 50 per cent target lies in the geographical mismatch between the demand for new housing and the supply of brownfield land. The areas where demand for housing is strongest (notably South East England) are those with the least amount of brownfield land.
- *Technical*: Although rates of reuse of brownfield land for housing have been impressive in recent years, Breheny questions whether such a rate can be maintained, arguing that the easier brownfield sites have already been redeveloped and the remaining sites will require large-scale subsidies before private sector developers will consider them.
- *Political*: Breheny quotes various sources of evidence to show how higher urban densities are likely to be resisted by many local authorities, in order to preserve well-established residential preferences. It will be crucial to ensure new brownfield developments are of a high quality and do not sacrifice standards of amenity if they are to be part of a viable long-term solution to urban regeneration.

In addition there are also a number of reasons – which could be termed 'social' – which will also need to be overcome. In recent years a significant amount of brownfield housing in large urban areas has been in the form of flats aimed at 'young professionals'. Examples are numerous and can be found in most of the UK's major cities (e.g. Salford Quays, London Docklands). To achieve an increase in the number of families moving into urban areas it will be necessary to overcome the negative perceptions associated with many of these areas. These include, fear of crime, poor air quality and poor-quality schools. Tackling these problems calls for a much more holistic approach to regeneration.

The government is seeking to address precisely these issues with the formation of an Urban Task Force. Chaired by Lord Rodgers, the Task

Force is co-ordinating research and policy development into making better use of brownfield land by identifying causes of urban decline in England and recommending practical solutions to bring people back into cities, towns and urban neighbourhoods. It is working alongside, advising and helping to develop initiatives through English Partnerships, local authorities and other key players. The Task Force's recommendations will have significant implications for the physical regeneration of urban areas across England.

Sustainable Transport

Emerging transport policies will also have implications for urban regeneration. The publication of PPG13 in 1994 (DoE, 1994) heralded the first practical step towards an integrated land-use transport policy which emphasises the role of public and non-polluting forms of transport and seeks to encourage patterns of development which reduce the need to travel by car. It is becoming increasingly important to demonstrate that steps have been taken to provide for access to new development by less polluting/ more energy efficient modes of travel. Overcoming such directives may favour development in highly accessible central urban areas, however, currently there remains a large amount of extant permissions granted prior to the introduction of PPG13 for developments in peripheral locations.

In the future, tougher restrictions on the use of the private car (particularly in congested urban areas experiencing poor air quality) such as restrictions on car parking and road-pricing could be a feature of central and local government policy. The extent of such policies and their potential effects on urban regeneration are as yet unknown and untested.

Sustainability and Economic Development

It is at the local level where many initiatives seeking to raise awareness and change individuals' behaviour are being developed in order to facilitate sustainable development. Much of this work is being undertaken through the Local Agenda 21 process by local authorities. Gibbs (1997) highlights the inability of many local authorities to integrate sustainable development objectives into their economic development and urban regeneration policies. In many respects regeneration and sustainability are seen as being in conflict, in that the pursuance of sustainable solutions may hinder economic growth. Gibbs suggests that the increasing involvement of the private sector in regeneration partnerships has further complicated this problem – one which commonly exists within local authorities. Though economic development strategies do increasingly acknowledge environmental objectives and constraints, they tend to focus on the physical environment rather than the wider 'global' issues. This is reflective of the guidance to which partnerships must conform when bidding for and implementing regeneration programmes such as the Single Regeneration Budget, City Challenge or European Structural Funds, all of which focus on economic outputs. Where environmental outputs are required these consist of hard physical improvements such as the amount of land reclaimed, number of trees planted etc.

Sustainability and the European Structural Funds

The need for sustainability to be addressed in major economic development programmes is, however, beginning to be addressed by the European Commission. Research carried out by ECOTEC (1997) for DGXVI identifies 16 areas of action which must be addressed by an Objective 2 programme if it is to help move the regional economy towards sustainability. These are classified into one of three steps or stages in moving a regional economy towards a model of sustainable development whereby economic growth continues whilst reducing the impact it has on the environment.

These steps are:

- business as usual;
- minimisation;
- the sustainable growth path.

Business as usual implies that all current environmental standards and regulations and procedures are complied with. Minimisation refers to development which seeks to go beyond existing standards by employing the best available technology. It also seeks to minimise environmental knock-on effects of new developments by directing them towards brownfield sites and locations where the necessary supporting infrastructure is already provided, i.e. close to existing public transport nodes. Though different localities will require differing patterns of development in order to move along a sustainable growth path, they will all share certain common features if they are to lay the basis for sustainability. These include:

- encouraging innovation in goods and services which use fewer environmental resources in use, manufacture or disposal;
- encouraging the economy to restructure towards these sectors;
- increasing the use of renewable energy;
- reducing the need to travel through use of IT, and integrated land-use and transport planning;
- the development and promotion of improved public transport.

In the Guidance for Programme Managers produced in association with the research, ECOTEC (1997) recommend a pragmatic approach to assessing whether the measures and associated actions proposed in a Single Programming Document will help to move the region towards sustainable development. This involves mapping the programme of measures (in turn) on to identified areas of action, and estimating the percentage of actions (or the budget) specified in the SPD which are devoted to each of the three steps towards sustainability. This takes the assessment of programming documents beyond the currently employed environmental appraisals by identifying specific actions which can both enhance sustainability and contribute towards economic growth and urban regeneration.

Employment Creation and the Environment

It will therefore become increasingly important that the profile of the environment is raised not simply as something worthy of protection but also as a potentially valuable source of new employment opportunities. Within the field of urban regeneration the potential for new employment opportunities is likely to be enhanced by the strategic objective for development on brownfield land. Public programmes to recover derelict urban sites will continue to be a feature of regeneration policy in the UK and the EU. Potential links with welfare to work programmes can provide both employment and training opportunities. The example offered by the Wise Group in Glasgow may prove to have widespread potential in this respect (see Box 5.10). Such initiatives will reinforce the link between environmental improvements and economic benefits.

New Visions of Urban Areas

The advent of the 'information economy' is dispersing production away from traditional urban locations. Different urban areas are responding in different ways. The largest (so-called 'world') cities such as London are developing a concentration of producer services. Smaller regional capitals (Manchester, Newcastle, Sheffield, etc.) are becoming centres for consumer services focusing on education, arts and cultural industries – exemplified by recent and proposed large regeneration projects such as Salford Quays, Manchester and Millennium Point, Birmingham. The impact of information and communication technologies on business services, leisure and retailing may well be even more significant for urban areas, if their purpose as centres for the location of such services becomes increasingly redundant. A major issue would then be: what will replace these activities to ensure the continued economic viability of many urban areas?

The changing structure of urban economies brings about requirements for new skill levels very different from those required when manufacturing was the dominant mode of production. This raises concerns that a requirement for more skilled employees (and fewer employees?) may result in large economically inactive urban populations, who are effectively excluded from the urban economy. These people have been termed the 'underclass'. Integrating these groups into urban society will be a major challenge for economic development in the new millennium.

Environmental pressures, particularly the drive towards more environmentally sustainable patterns of development, will also impact on the future development of cities. At a strategic level it may focus development activity on existing urban areas (i.e. brownfield land) in order to avoid developing greenfield sites, and encourage concentrations of activity in these areas (the notion of the compact city). Over the next few decades this will be exemplified by the need to accommodate the majority of the demand for new housing in the UK in existing urban areas. Achieving this may well require significant government intervention in terms of land preparation and reclamation, financial incentives and land-use regulation.

The focus on sustainable development will also have significant implications for infrastructure provision, particularly transport. Moves to restrict the use of the private car in urban areas, including road-pricing will have significant, if unpredictable, effects on land-use patterns and investment in our cities. There is already a move towards encouraging the use of public transport for urban journeys, even if this has not yet been translated into hard infrastructure provision other than in a few exceptional examples (e.g. light rail schemes in Sheffield and Manchester).

Key Issues and Actions

- Be clear about the spatial scale and the time scale of the proposed renewal actions.
- Understand the ownership and the economic/market trends affecting the physical stock.
- Be clear about the role of the physical stock in the renewal strategy – lead/flagship/supply side; 'enabling' or responding to demand; integrated.
- Undertake SWOT analysis of stock.
- Develop a clear vision and a strategic design for the renewed physical conditions.
- Ensure this fits the emerging role of the area, integrates with the other dimensions of renewal and is developed with the participation of the appropriate partners in the area.
- Establish institutional mechanisms for implementation and continued maintenance of the schemes.
- Establish mechanisms for capital, operation and maintenance funding.
- Understand the economic rationale for environmental improvements.
- Ensure the approaches can respond to changing government delivery strategies, and changing social and economic trends.

References

Breheny, M. (1997) Urban compaction: feasible and acceptable? *Cities*, Vol. 14, no. 4, pp. 209–17.

Department of the Environment (DoE) (1994) *Planning Policy Guidance Note Transport*, HMSO, London.

Department of the Environment, Transport and the Regions (DETR) (1998) *Planning for the Communities of the Future*, HMSO, London.

ECOTEC (1988) *Improving Urban Areas*, HMSO, London.

ECOTEC (1996) *The Economic Impacts of Canal Development Schemes*. A Report to British Waterways, ECOTEC, Birmingham.

ECOTEC (1997) *Encouraging Sustainable Development through Objective 2 Programmes: guidance for Programme Managers*. A Report to DGXVI of the European Commission, ECOTEC, Birmingham.

Gibbs, D. (1997) Urban sustainability and economic development in the United Kingdom: exploring the contradictions, *Cities*, Vol. 14, no. 4, pp. 203–8.

Healey, P. (1995) The institutional challenge for sustainable urban regeneration, *Cities*, Vol. 12, no. 4, pp. 221–30.

Loftman, P. and Nevin, B. (1996) Going for growth: prestige projects in three British cities, *Urban Studies*, Vol. 33 no. 6, pp. 991–1019.

Roberts, P. (1995) *Environmentally Sustainable Business*, Paul Chapman, London.

Turok, I. (1992) Property-led urban regeneration: panacea or placebo, *Environment and Planning A*, Vol. 24, no. 3, pp. 361–79.

UK Government (1996) *Household Growth: Where Shall We Live?* Cmnd 3471, HMSO, London.

Further Reading

Department of the Environment (1995) *The Impact of Environmental Improvements on Urban Regeneration*, HMSO, London.

Department of the Environment (1997) *Bidding Guidance: A Guide to Bidding for Resources from the Government's Single Regeneration Budget Challenge Fund (Round 4)*, DoE, London.

English Partnerships (1995) *Investment Guide*, English Partnerships, London. Includes criteria against which EP assess bids for resources.

English Partnerships (1996) *Working With Our Partners: A Guide to Sources of Funding for Regeneration Projects*, English Partnerships, London. The first edition is now out of print but provides an brief introduction to the variety of activity undertaken by EP and an excellent overview of the range of sources of funding for regeneration. An updated second edition is expected.

European Commission (1997) *Urban Pilot Project: Annual Report 1996*, Provides an insight into the range of regeneration projects funded by this Community Initiative, Office for Official Publications of the European Communities, Luxembourg.

HM Treasury (1995) *A Framework for the Evaluation of Regeneration Projects and Programmes*, HMSO, London.

Llewelyn Davies (1996) *The Re-use of Brownfield Land for Housing*, Joseph Rowntree Foundation, York.

Robertson, D. and Bailey, N. (1995) Housing renewal and urban regeneration, *Housing Research Review*, Vol. 8, pp. 10–14.

6 Social and Community Issues

Brian Jacobs and Clive Dutton

Introduction

This chapter provides an overview of some important issues concerning the role of local communities in urban regeneration. The boxed case studies refer to differing experiences. They provide practical examples of how policies cater for community needs and encourage community involvement in urban affairs. The case of the Tipton Challenge Partnership in Sandwell in the West Midlands features prominently as an example of good practice. Tipton Challenge provides a refreshing example of a successful and innovative initiative facilitating the participation of local people in community affairs. The Tipton case, and other illustrations, show how programmes and projects require effective and efficient management if they are to cater for a broad range of community needs. The chapter thus covers some important interrelated concerns that arise when managers of urban regeneration initiatives deal with diverse local issues and needs. Local authorities, company sponsors and voluntary organisations have to ensure that programmes benefit local people and produce value for money. The topics here serve to identify considerations taken into account when implementing local programmes. The topics, reviewed here during 1997, concern:

- the definition of 'community';
- the needs of local people in communities and the provisions to meet those needs;
- the special needs of community groups;
- the development of shared goals to promote the improvement of economic and social conditions;
- representativeness in communities;
- community empowerment;
- effective community partnerships;
- capacity-building in local communities;
- lessons from good practice.

Even though the creation of employment is at the top of the list of priorities of many community-based regeneration schemes, the emphasis in this chapter is on a wider list of issues. However, recognising the importance of employment, this topic is dealt with in detail in Chapter 7.

Defining Community

Involving Communities in Urban and Rural Regeneration, produced by the Department of the Environment, Transport and the Regions (DETR, 1997a), provides a working policy definition of 'community' as the people working and living in defined areas covered by regeneration programmes. This policy definition envisages 'communities' as enclosed by the boundaries set by the Single Regeneration Budget Challenge Fund (SRBCF), City Challenge, the Task Forces and Housing Action Trusts. According to the DETR, the private and voluntary sectors and local authorities try to identify the most immediately felt problems of people within the contexts provided by such communities. However, this view is very restrictive, so the DETR guide usefully expands the definition of 'community' by referring to criteria developed by the Local Government Management Board (LGMB). Box 6.1 presents the LGMB criteria.

Box 6.1   The characteristics of communities

Communities may be defined by reference, for example, to their:

- personal attributes (such as age, gender, ethnicity, kinship);
- beliefs (stemming from religious, cultural or political values);
- economic position (occupational or employment status, income or wealth, housing tenure);
- skills (educational experience, professional qualifications);
- relationship to local services (tenants, patients, carers, providers);
- place (attachments to neighbourhood, village, city or nation).

Source: DETR (1997a), quoting the Local Government Management Board.

The LGMB approach allows for widely different 'communities' based on a rich variety of local experiences and perceptions. One or more of the characteristics in Box 6.1 combine to create particular community identities and attitudes. This implies that the conception of community varies to such an extent that narrow technical definitions are unsatisfactory.

Disagreements over the meaning of 'community' therefore abound, but the generally favourable images and associations that the term community provides, forms a basis for partnership and innovation. Communities can then support economic activities that benefit local people (Burns, Hambleton and Hoggett, 1994) especially when they engender powerful emotive connotations that derive from a sense of togetherness and social identity. Notions of solidarity, pride and identity tend to bind people together (Tilly, 1974). Consequently, the most viable and ultimately successful communities are those that engender a sense of belonging and partnership between people. According to Charles Handy (1997) vibrant communities evoke trust between people and form civil associations that contribute to local welfare. Local people establish supportive relationships producing a sense of togetherness that is not constrained by physical boundaries. For Handy, people develop rules and values that enable individuals to work together and prosper.

Community Needs and Provisions

Unfortunately, the conditions for the establishment of viable and successful communities are not always present. An important recent study by the United Nations and the World Bank (World Resources Institute, 1996) indicates the problems facing urban policy-makers world-wide. Policymakers have to deal with complex and interrelated problems including shortages of funding for local amenities, overburdened social and education provisions and the lack of access of citizens to affordable health services. The study shows that in every nation citizen involvement is vital to ensure the success of public policies and the prosperity of communities.

In the United Kingdom, the specific nature of the problems facing urban communities differs from those in developing countries, but local groups here are as keen as elsewhere to improve the quality of life. In this country, there is an established tradition of community and voluntary effort that provides a strong foundation for urban regeneration. Geddes (1995) argues that for communities the challenge is to improve their access, extend social and economic opportunities and develop local services to become more effective in meeting local needs. In practice, this has led to policy actions to deal with policy areas that increasingly have to be viewed together. The main policy areas are mentioned in the SRBCF Round 4 guidance shown in Box 6.2.

The policy concerns in Box 6.2 require co-ordinated action and funding because they together influence the social and economic prospects of communities. In the case studies below, we select health as a particularly useful focus because it underlines the interconnections between different policy areas. For example, with an ageing population in Britain there is an increasing demand for health care and community support. Good health depends upon good housing, adequate social provisions, a pleasant environment and leisure, sport and recreation opportunities. Such thinking lies behind the government's decision to promote Health Action Zones and related regeneration programmes (DETR, 1997c).

The good health of communities improves the quality of life of local people and develops a greater pride in localities. When people enjoy a better life, it gives them the confidence to plan for the future within a pleasant community context. Health initiatives can also significantly enhance the capacity of localities to provide locally responsive services through the extension of relevant skills and capabilities. In health, alternatives to more traditional service provisions therefore provide exciting

Box 6.2   Related community policy concerns: Single Regeneration Budget Challenge Fund Round 4 – Supplementary Guidance, 1997

- Welfare to work. Challenge Fund schemes have an important role to play in the government's new measures to tackle unemployment, educational attainment and social deprivation. When possible schemes should complement new opportunities/programmes which will soon be available under the Welfare to Work initiative and in Employment Zones.
- Education Action Zones. Government Offices/the Employment Service will supply further information to partnerships as the detail of these developments becomes clear.
- Capital receipts. Challenge Fund schemes can effectively complement measures to tackle housing need or promote housing-related regeneration which is proposed for support from the release of housing capital receipts.
- Crime. The government intends, in the Crime and Disorder Bill, to give local authorities and the police a joint responsibility to develop local partnerships to tackle and prevent crime in consultation with the local community. The Challenge Fund can help local partnerships draw up and implement strategies for tackling crime. The new duty will not be in place by the closing date of the bidding round, but bidders should seek to ensure that any bids with a significant crime prevention or community safety element are compatible with such local plans as may have already been drawn up by the relevant local and police authorities. Re-establishing a sense of community and building capacity in the voluntary sector is also a government priority which the Challenge Fund may be able to help.
- Drugs. Tackling drugs requires co-ordinated action at every level. The government is committed to appointing a 'Drugs Tsar' to co-ordinate action nationally, regionally and locally. The SRB Challenge Fund provides an opportunity for local services and local people to work together through Drug Action Teams to tackle the problem within communities. The Prime Minister has already made clear that regeneration work is central to the fight against drugs.
- Ethnic minorities. The Crime and Disorder Bill will also include specific offences to tackle racial violence and harassment. Challenge Fund schemes, especially those which help develop a sense of community or build capacity in the voluntary sector and those involving crime prevention, community safety and victim support, can complement other work to tackle racial violence and harassment in local communities, and also target economic development and training initiatives on such communities.
- Public health. The government is developing a new health strategy which will recognise the impact that social conditions such as poverty, poor housing, unemployment and a polluted environment have on people's health. Challenge Fund schemes can contribute to improvements in public health through these conditions, as well as by developing partnerships with local health bodies to promote healthier lifestyles and improve access to community-based health facilities.
- Vulnerable groups. The government's commitment to tackle social exclusion will have particular relevance for vulnerable groups in the community such as homeless people, frail elderly people and those with mental illness. Challenge Fund schemes can develop targeted housing projects to promote community-based care and greater quality of life for these groups.

In addition, Challenge Fund schemes should complement wherever possible other programmes, for example, capital investment in renewal of school buildings and the development of information technology.

Source: DETR, 1997b.

opportunities for the development of new approaches to service delivery and personal development. However, this can only be achieved if policies cater for the real needs of communities.

The Special Needs of Community Groups

John Bennington (1994) points to the exclusion of groups based on race, gender and age. These are groups with special needs arising from various forms of discrimination. Discrimination against ethnic minority and socially deprived groups reinforces the problems faced by people in communities and underscores their problems. It is often difficult to reconcile differing needs when localities accommodate many competing interests, values and social groups. Policy-makers therefore try to promote some sense of inclusion and belonging. Policy goals frequently expand themes that allow for gender, religious faith, race and age differences as well as how to overcome disadvantage. Policy objectives increasingly identify specific groups, such as the elderly and women, so that groups can work effectively together through appropriate consultative frameworks. The SRBCF encourages bidders for funding to base their cases for support on specified needs identified in their target areas, and regional frameworks will extend the recognition of need more broadly allowing for a more comprehensive set of policy initiatives to deal with exclusion. Box 6.3 indicates how ethnic minority organisations can become directly involved in identifying needs within the community.

Box 6.3   Ethnic minority needs in Sandwell

- Following its Round 1 SRB bid, Sandwell Regeneration Partnership held a conference aimed at the involvement of the ethnic minority community and the voluntary sector in the Partnership's second round SRB bid, Sandwell Capacity Building for Urban Regeneration.
- Black-led community organisations were invited to submit project proposals to the partnership. Once these had been received, the partnership attempted to match up those organisations who had submitted bids for similar projects. A series of workshops was then held giving community groups an opportunity to discuss their proposals. The value of project proposals far exceeded the likely level of available resources. The project proposals were refined and resubmitted to the partnership. Surgeries were held with unsuccessful bidders to discuss alternative funding mechanisms.
- The partnership's successful Round 2 SRB bid was driven by the Bid Development Group; a body consisting of 40 key members of the local community, including representatives of ethnic minority communities. Some 30 per cent of the £7.1 million to be allocated over a seven-year period is targeted specifically at the borough's ethnic minority communities.

Source: DETR (1997a).

A Shared Vision

Disadvantaged groups have interests in the improvement of localities and they all have ideas to contribute on how communities can be regenerated. However, tensions often exist between conflicting local interests, racial and ethnic groups or opposing environmental lobbies. It is therefore important to achieve a vision for the community that can be shared by all and which defines priorities for action.

The involvement of local people working with the public and private sectors improves the quality of policy decisions and secures the more effective implementation of local programmes. Burns, Hambleton and Hoggett (1994) show how local councillors, public officials, company executives and citizens support the community initiatives that bring groups together to produce a wider awareness of how local problems can be overcome. Burns, Hambleton and Hoggett (1994) refer to the unifying impulse that community identity can provide. The unifying impulse of communities provides a firm basis for linking different interests and establishing the shared community vision. Communities can foster co-operation, common purpose and confidence (CBI, 1988; Taub, 1994) and reflect these in vision statements and policy commitments. The success of communities will depend upon the degree to which they can cohere around aims and objectives, and the degree to which local people can identify with those objectives.

Representativeness

Despite the importance of including a wide range of interests, active groups and individuals that appear to be influential in a locality are not necessarily representative of local people or businesses. Frequently, groups or organisations do achieve a respected standing within a community based upon their association with the area and their skills and expertise. Independent or voluntary sector organisations can provide important services within communities (Ware, 1989) and they often represent groups that find it difficult to speak independently. Voluntary organisations provide professional assistance and advice and other assistance to groups seeking funding from public and private sources. Voluntary organisations can work with local people to articulate local needs and improve the quality of management in community projects.

The DETR (1997a) guide refers to the importance of creating representative boards to run local programmes. The guide suggests that 'credible' representatives should represent networks of local groups. It is then possible to 'empower' communities through the widest and most effective representation of local interests. The composition of boards, the election or selection of representatives and agreement on appropriate structures present hard choices. Representativeness and community access to local authorities and partnerships can be difficult to achieve if a limited number of groups predominate over others. Where groups develop strong identities, public policies tend to respond to the demands of the leading representatives of the groups concerned. Conflicts can arise especially when excluded community leaders challenge the assumptions that underpin prevailing public policies. Individual or sectional interests can then conflict with the interests of the community as a whole. Representativeness in community initiatives should therefore establish the 'ownership' of initiatives by the community (Farnell *et al.*, 1994) so that there can be as broad as possible acceptance of the goals pursued.

Empowerment

Empowerment extends the 'ownership' of programmes and projects to communities, and provides local people with responsibilities for influencing and taking decisions on SRBCF management boards and so on. Another popular mechanism for empowerment is through the development of local enterprises that employ people so that they can gain from their participation in the economy and develop skills for the future. Support for economic initiatives in regenerating communities finds expression in the policy objectives of governments and in the prescriptions of the private sector (BITC, 1986; CBI, 1988). In the United States, many programmes have included community enterprise as a central strategy for neighbourhood regeneration and empowerment (Taub, 1994). In Britain, politicians from all the political parties support market solutions in urban regeneration and embrace private initiative, public–private partnerships, small businesses and entrepreneurship.

Empowerment requires policies that enable citizens to gain greater access to services and to have more say on the use of community resources such as housing. Also, 'community development' (West, 1993) interprets empowerment as a practical policy strategy. Local authorities and the private sector can play an important role in assisting people to empower themselves and to make informed choices concerning their futures. For example, policies can provide tenants in public housing with the chance to own their homes, give parents greater choice between schools, and give patients the opportunity to make selections about the care they receive. Diversity, choice and enterprise are thus firmly established in the vocabulary of community regeneration because they are essential ingredients of empowerment.

Effective Regeneration Partnerships

As discussed in Chapter 3, partnerships are the organisational vehicles of community regeneration and empowerment. The SRBCF uses community partnership as a way of harnessing the talents and resources of communities in urban regeneration. Partnerships combine the efforts of three broadly defined groups of organisation:

- Public organisations involved in urban regeneration include local authorities, central government departments, the Government Offices for the Regions, National Health Service Trusts, and the Police. There are also other agencies such as the business-led Training and Enterprise Councils that play a prominent part in urban regeneration.
- For-profit private sector organisations, such as companies and financial institutions, play an increasing role in communities through corporate social responsibility programmes and sponsorship. For example, Business in the Community (BITC) is a non-profit organisation that brings companies together with voluntary organisations and local government to implement programmes at the local level.
- Not for profit private sector organisations come in many forms. The National Council for Voluntary Organisations (NCVO, 1996a) provides a detailed annual breakdown of the voluntary sector. Practitioners often subdivide groups as private non-profit organisations, voluntary organisations, and charities. But, however we define groups, they all make significant contributions to the betterment of communities and to the development of effective partnerships.

As the NCVO policy suggests, partnerships need to provide effective management structures to meet the policy aims and objectives agreed by partners. Companies and local authorities also advocate effective partnership management.

- Partnerships should enable community programme managers to work alongside politicians, public officials and representatives of the private corporate and non-profit sectors. The co-ordination of different interests is sometimes difficult, but is important to ensure the success of programmes.
- Partnerships should provide effective management structures for the implementation of the shared policy goals of partners. Management and organisational structures need to be appropriate to the objectives set by managers of local initiatives.
- Partnerships should link different programmes (such as the SRBCF, European initiatives, TEC programmes and private initiatives) into coherent strategic plans (see also Chapter 11 on organisation and management). Measures can help to establish a wider sense of community as well as providing new opportunities in social, health, education and recreational provision. In partnership, local health and education authorities, city councils and voluntary organisations can provide information to families, enhance access to services and promote enhanced opportunities.

Box 6.4   *The role of voluntary organisations*

The importance of the voluntary sector in urban regeneration is evidenced by the existence of over 125,000 voluntary community organisations. An independent Commission on the Future of the Voluntary Sector supported and initiated by the NCVO and chaired by Professor Nicholas Deakin, consulted extensively with voluntary organisations to assess their role. The commission presented its report entitled 'Meeting the challenge of change voluntary action into the 21st century' in 1996. This report found that: 'voluntary and community organisations in all their diversity are a major resource. Their independence must be safeguarded.' Further, the NCVO stated that the sector should be governed by six basic principles:

- Public policy needs to recognise the unique qualities of voluntary action.
- Partnership must be on an equal basis.
- The role of users is crucial to the sector.
- Voluntary bodies must always be free to act as advocates.
- The sector must be managed professionally, without deflecting from its purposes and aims.
- Diversity of funding sources is one of the best guarantees of independence.

This led to a number of actions to be taken by the NCVO and voluntary organisations as follows:

- supporting a concordat or code of good practice on future relations between the government and the voluntary sector, encouraging government to adopt a strategic approach to its relations with the sector;
- encouraging a change in the legal definition of charity to one based on a single definition of public benefit, enabling more organisations to gain charitable status;
- building on NCVO's previous work with the Charity Commission, on the Nolan Committee and with the publication of the *Good Trustee Guide* to develop models of performance and governance for voluntary organisations;
- building on the Fiscal Working Party in liaison with Business in the Community, the Institute of Charity Fundraising Managers and Charities Aid Foundation to create a consortium to investigate new sources of funding;
- promoting Europe not just as a source of funds but also as a means of developing transnational partnerships;
- developing quality standards for the sector to demonstrate good governance and effectiveness;
- developing and supporting links with business through board membership, gifts in kind and skill sharing.

Source: NCVO (1996b).

Some writers have alluded to the limitations of the partnership model. Farnell *et al.* (1994) describe the American-inspired broad-based organising (BBO) where local 'actions' promote initiatives that are fully accountable to communities. The BBO approach places emphasis upon the exertion of 'political pressure' through bottom-up 'organisation-building' (Farnell *et al.*, 1994) rather than by way of what critics regard as the 'top-down' approach of partnerships. The BBO approach implies a broad-based form of community involvement through 'day-to-day activism' that encourages the involvement of 'the hitherto voiceless' in communities (Farnell *et al.*, 1994). However, it is difficult to achieve a genuinely bottom-up structure in practice in Britain especially when so many local groups lack the capabilities effectively to involve themselves in urban regeneration initiatives. The capacities of communities need to be substantially strengthened as a result.

Capacity-building

When community organisations gain access to the policy process through partnerships or through other means, they need to develop their capacities to engage in local economic development and social initiatives. The DETR (1997a) recognises the role of communities in helping with the strategic direction and management of programmes and the consequent need for communities to develop the capacity to achieve this. For the DETR, capacity-building is about:

- skills: project planning, budgeting and fund-raising, management, organisation, development, brokerage and networking;
- knowledge: of the programmes and institutions of regeneration, their systems, priorities, key personnel;
- resources: essential if local organisations are to be able to get things done;
- power and influence: the ability to exert influence over the plans, priorities and actions of key local (and national) agencies.

Capacity-building is a process that enhances:

- the empowerment of communities because people increasingly do things for themselves;
- the ability to create structures and networks to assist this process;
- skills to enable local people to take charge of their futures.

Case Studies: Social and Health Issues

The inclusion of social, welfare and health objectives in urban regeneration signifies the creation of new social infrastructures that clearly demonstrate the need for the development of such capacities. Local initiatives concentrate on the improvement of the health of local people and the provision of social support services that imply a requirement for local expertise across a range of activities.

Health criteria are now included in many SRBCF schemes. This marks a stronger emphasis on social regeneration to complement physical regeneration. A 1996 British Urban Regeneration conference (Health of the nation: community regeneration perspectives) indicated the relationships between social and health issues such as employment and housing (Harcup, 1996). The development and improvement of the social infrastructures of communities was a major theme at the 1996 United Nations Conference on Human Settlements. Good health and environment are key indicators of a good quality of life, so cities should 'embody the diversity and energy of human pursuits' and act as the 'engines of social progress' (World Resources Institute, 1996, p. ix). The European Commission's Social Action Programme will promote new initiatives on health provision and the improvement of information systems for communities on health-related issues. The Commission's strategy concentrates upon new mechanisms for providing health care at the local level and the development of integrated policies linking health to related social and economic problems. In the United States, Empowerment Zones and Enterprise Communities often focus upon social and health issues that stress the involvement of local people in initiatives. The EZ/EC initiative places emphasis upon distressed communities developing their capacities for self-improvement, innovation and creativity (HUD, 1995). Imbroscio *et al.* (1995) refer to the role of 'human investment' programmes in central-city regeneration in the USA. Human investment seeks to make people more productive as members of society and involves the development of skills in health, education and training.

In the north-west of England, Bolton City Challenge has developed a strategy that includes projects involving Bolton Health Authority and the Health for Bolton scheme in partnership with Bolton social services. The approach builds the capacity of the community to provide support services through effective management. Health promotion and health awareness are of central importance as they complement sport, leisure and environmental improvements. The initiative includes drop-in access to local people through a Health Information Centre. The partnership enlists the support of midwives in expanding maternity services to ethnic minority women. There is also a drugs and substance misuse scheme and a mental health project (Bolton City Challenge, 1995).

Integrated policies have been recognised by the European Commission. For example, an exchange programme established between 11 European participants concerned policies in urban areas aimed at socially disadvantaged groups. The 'European Network Eleven' shared experiences and disseminated information about good practice. Birmingham, Luton and Stoke-on-Trent joined in the programme along with Berlin, Rome, Naples, Lyons, Tours, Amiens, Helsinki and Regio Calabria. The exchange provided the opportunity to create a wider European network for the dissemination of information on policy and practical experiences as well as the possibility of joint funding and projects linked to European Objective 2 initiatives. In Network Eleven cities, officials recognised the linkages between programmes such as the European Regional Development Fund, EU Community Initiatives for run-down mining areas, Regional Challenge and the SRBCF.

Box 6.5   Bolton City Challenge

By Susan Richardson (Bolton City Challenge)

Quality of life is central to Bolton City Challenge's remit to regenerate the Halliwell area of the town.

Health and community issues are held in the same esteem and importance as the economic regeneration of this area with its high unemployment rate and associated social problems. But it is not just physical illness that City Challenge projects tackle. It also:

- addresses the effects on mental health of the stresses of modern day living through the Mental Health Initiative;
- provides a dedicated multi-disciplinary team that works to provide a drop-in centre, which has more than 230 members;
- supported a housing scheme and an employment service to benefit those with mental health problems and their carers.

The ethos has been to listen to local people and come up with health solutions that fit their needs.

Another aspect of City Challenge's work is supporting local organisations through its Community Fund which distributes grants to a diverse range of groups.

For example, there is Shopmobility which provides wheelchairs for disabled people to borrow for all sorts of social activities. The scheme recently came second in the community section of the NHS Health Challenge Awards and has 700 users.

The Betty Hamer Life Education Unit received £60,000 from the City Challenge Community Fund. This is a mobile classroom that visits all primary schools in the area, educating children about health through interactive activities.

Halliwell's youth are also part of environmental activities, such as the Mean Green Team which is a group of local youngsters who organise and participate in local clean-ups, tree planting and a whole host of other activities.

The Recycling Factory is a joint community enterprise initiative between Bolton City Challenge and Lancashire Wild Life Trust. It encourages local businesses and schools to save aluminium cans which will then be collected and recycled with resultant funding reinvested into further enterprise initiatives.

City Challenge has also helped to establish Crossover '97. This is a unique initiative bringing together a number of local organisations including Bolton Metro's Youth and Leisure Services, Community Healthcare and NACRO. They have achieved considerable success with activity programmes aimed at those between 5 to 25 years with a wide range of schemes such as drama workshops and football courses organised in partnership with the town's nationally renowned theatre, The Octagon, and Premier football club, Bolton Wanderers.

The City Challenge philosophy is that investment should be put to work to support projects run by the people for their community as evidence indicates social and economic improvements cannot be achieved simply by spending money.

City Challenge has spearheaded that attitude and established a number of community-level initiatives that are now being adopted by other areas. An example of its approach is Health Link which offers medical advice in an informal 'shop' atmosphere designed to break down the community's reluctance to use traditional practices.

One Health Link service that has been extremely successful is the Cardiac Rehabilitation service. It involves one to one visits to patients by a qualified nursing sister on their discharge from hospital to give them and their families advice and to help them avoid the risk of further coronary problems.

Source: Written communication from Susan Richardson.

Case Study: The Tipton Experience

The Tipton Challenge Partnership in the West Midlands provides an important example of how urban regeneration can be about the creation of new community social provisions that provide direct access to local people and enhance community capacity building. The case shows the combination of actions designed to involve the community. The main activities of the partnership are shown in Box 6.6. The account below was written in late 1997, during the lifetime of the partnership.

Box 6.6   The Tipton Challenge Partnership main activities

- Improving access to employment through education and training.
- Preserving existing business and stimulating new businesses and enterprise.
- Improving housing conditions and offering a wider choice of housing.
- Reducing crime and enhancing community health and safety.
- Improving image and environmental quality with the removal of dereliction.
- Providing new and enhancing existing sports, leisure and community facilities.
- Sharing the vision.

Source: adapted from Tipton Challenge Partnership (1997).

It is the substantial development of a health-oriented approach that makes the regeneration of Tipton of special interest. The particular interest lies in the pivotal role that the health sector has played in the Tipton story. Whereas economic factors tend normally to take centre-stage in so many regeneration efforts, in Tipton's case the shared focus has been very much on improving health and quality of life. In this way, health is seen as a catalyst for the future economic prosperity of the town.

Planning for real, public consultation, public information and participation exercises related to development or individual initiatives are not new. They have been used to varying degrees for over 20 years. Their effectiveness is varied and in many cases they were undertaken to enable a particular 'box' to be ticked on a list of tasks that ought to be undertaken by the promoter or regulator of the scheme involved. Often such involvement occurred relatively late in the redevelopment process when economic, property, commercial and professional issues or preferences had been largely determined. There have been notable exceptions, but that was largely the picture. This resulted in schemes that could not be sustained because they alienated the very people who were required to live, shop, visit or work in an area where major transformation was planned; or more precisely, the community.

The 1990s brought significant changes. City Challenge introduced its process focus to the implementation of 31 time-limited area-based initiatives. Such an approach is endorsed and encouraged in SRB schemes. The government guidance, in 1991, to the 20 authorities invited to bid for City Challenge status was that 'partnerships' would be a prerequisite, led by the local authority but incorporating significantly the private sector, other public sector agencies and the local community. The regenerative propositions were aimed at both geographic areas (the average being in the region of 3 square miles and 25,000 residents) and to address comprehensively the revitalisation of an area (the social as well as physical and economic issues). This was a significant change to both the emphasis and approach to urban policy.

However, communities, particularly those that have suffered in the demoralisation and demotivation of a third of a century of decline, cannot readily engage in 'partnership' working as equal players with the other sectors without preparation and ongoing support (both financial and nonfinancial). To illustrate this, examples are taken from Sandwell in the Black Country in the West Midlands. This is a particularly deprived borough; the third most deprived outside London and the ninth in England.

The first example relates to a community development 'pilot' project initiated by Sandwell Metropolitan Borough Council (MBC) in 1989 in Tipton; one of the six urban villages and towns that make up the borough. Tipton, an urban district council up until 1966, was subsumed by larger municipal administrations such as the West Midlands Metropolitan County and later by the Metropolitan District. In parallel with periods of municipal change and loss of autonomy for local issues came decline through successive recessions. Full employment gave way to 20 per cent unemployment in little over 15 years. Tipton not only had its heavy manufacturing base dismantled, but also lost its former identity and opportunity to manage and influence its local affairs.

The community pilot project was selected for Tipton for a variety of reasons:

- Crucial was the fact that the area had not benefited from any previous significant urban programme initiative.
- The project involved four community development workers, employed by the local authority working full time within the communities of Tipton.
- Their goal was to help local voluntary and interest groups to focus upon issues collectively and to develop a greater understanding and contact with local authority service providers.

The initiative created an 'umbrella' organisation of local voluntary sector groups, named the Tipton Action Group. The initiative, however, did not bring with it substantial funds to support the group. More importantly, the funds did not enable any projects to be undertaken to either tackle local issues individually or to address poverty on a wider scale.

When the government announced its second and, as it turned out, last City Challenge bidding round, a proposal for Tipton appeared compelling.

Not only was the area in desperate need of regeneration, with opportunities that, if realised, could help address some of the problems, but there was already a community mobilised, focused and inducted to interface with the public sector. It enabled a community-based bid to be formulated to address local issues and maximise the employment opportunities created by the nearby Black Country Development Corporation.

The government requirement for partnerships can often provide a dilemma as to the character and method of engaging local people in the process of regeneration. The token involvement of people can be a danger, particularly where there is pressure to establish partnerships at very short notice. In Tipton, to a large degree, the situation was different, as the Tipton Action Group could claim to be representative. It could genuinely provide a mechanism for nominating its own representatives to the Partnership Board and a plethora of subgroups dealing with topic-based policy, implementation and monitoring responsibilities. Box 6.7 shows the make up of the delivery mechanism for Tipton Challenge Partnership, the organisation formed to orchestrate the City Challenge programme.

Box 6.7   Tipton Challenge Partnership: capacity-building through participation

Community representatives are nominated and elected by the various neighbourhood groups overseen by Tipton Action Group, the Muslim community and the Tipton Young People's Forum. The latter represents people between the ages of 11–25; the elections involved proportionately more people than vote in Tipton at local government elections. The Chair of the forum has a place on the board. In each case, the community determines the nomination and the term and the right to replace. As a result, the largest representative group on the Challenge Board is the community.

The Tipton Action Group Board members help oversee a £174 million regeneration programme, and they are responsible for administering a community block fund programme. This involves £150,000 per year available for local community organisations by application.

Through preparing communities in advance of the regeneration initiative, the community can readily become engaged as a representative, responsible, talented and committed contributor to the regeneration of the area. It may be asked, how could it be any other way? Regeneration schemes come and go, but the community, by whatever definition, goes on forever.

The Tipton Health Strategy

Tailoring local circumstances with a community dimension can provide surprising results if the partnership involved is flexible, creative and alert to opportunities linked to local circumstances. A good example of this is how the Tipton Challenge Partnership has developed a health strategy to both address core symptoms of poverty and turn it to regenerative advantage.

Box 6.8   Building on the Tipton experience

The community development workers in Sandwell moved to another town a
year after Challenge started to undertake similar 'capacity-building' work in
Cradley Heath, another area of the borough which had hitherto received little
in the way of urban funded initiatives, but where a large estates action project
was about to start.

The real lessons learnt in 'investing in the community' in this way have been
developed by Sandwell MBC a stage further and impressively have been rewarded for its approach by the government in the form of SRB Round 2
funding. The lessons learnt from Tipton are that true regeneration, which is to
be long-lasting, must involve local people from the start in an area's regeneration. They must be assisted before, during and after the particular initiative.
The investment in preparing communities to be full, responsible, and accountable contributors is extremely important. The SRB Round 2 initiative is solely
related to the process of empowering and enabling local communities in the
regeneration of their neighbourhoods, their localities and their borough. During seven years, £7 million has been awarded by government to assist this;
recognising that the process is important to the character, the quality and the
durability of the product. Investing in communities in this way is extremely
significant. This approach underscores the significance of process especially in
the way that the public sector views and approaches the way it delivers local
services. The mechanism by which the local urban voices can assist public
sector agencies in how they best meet local needs and tackle local issues has
much to learn from the Tipton, and other City Challenge and SRB, models.

Similarly, this approach can help in ensuring that regenerative solutions to
deep-rooted social, economic and physical problems are tailored to the circumstances of the individual area. Whilst much can be learnt from what other
practitioners have done (both right and wrong) in other parts of the United
Kingdom and abroad, there can be no template, blueprint or prescription for a
particular town or neighbourhood. All have their own special characteristics
and strengths, needs and opportunities. Each has its own community characteristics with different special needs.

Tipton displayed all the characteristics of communities that developed
rapidly during the Industrial Revolution but whose social and economic
conditions are now among the worst in the country. Life expectancy was
low – the average life expectancy in Tipton was seven years below the
national average. Health was bad. There was a much higher than average
incidence of the illnesses, such as coronary heart disease, which shorten life
and increase disability. Housing was poor – 22 per cent of homes were
classified as unfit to live in. Wages were low – 75 per cent of residents had
incomes of less than £100 per week. Long-term unemployment was high;
running at 45 per cent of the workforce out of work for more than six
months.

Three years on from winning its City Challenge bid, Tipton is beginning to
look different and feel different. Some 400 projects are being undertaken – in
education, training, housing, environmental improvement, land reclamation,
job creation, leisure, community safety and health. All these projects aim to
enhance the quality of life of Tipton residents. They also reflect the priorities

of the people themselves. Directly involving the local community has been a guiding principle throughout. The projects have flowed from community needs. They have not been superimposed from outside.

Already Tipton is seeing tangible benefits. Unemployment has dropped by 50 per cent since City Challenge started. The spirit of Tipton has been recharged and revitalised. City Challenge has had many spin-offs for Tipton. A significant number of the projects now under way are health related including a mobile community health team, a food co-operative to make healthy produce available at affordable prices and healthy lifestyle advice for parents of local primary school children.

Box 6.9   Tipton Challenge Partnership: the Neptune Health Park

The Neptune Health Park grew out of a number of coincidental factors:

- the determination of doctors in a go-ahead general medical practice in Tipton to improve their premises and provide a better quality of service to their patients;
- the determination of Sandwell's major provider of hospital and community health services to ensure that it is meeting the real needs of Tipton people;
- the strong commitment of Sandwell's health authorities to tackling the causes of disease and reducing the inequalities between different parts of the borough;
- the growing bond of partnership between local health services and Sandwell Metropolitan Borough Council;
- the flexible thinking stimulated by the success of City Challenge.

The overriding desire of all parties involved to put health at the top of the town's agenda led to the Neptune Health Park idea. Out of a number of vital exchanges came Neptune Health Park. Side by side on a 5-acre site in the centre of Tipton would be a whole range of new health facilities. As well as the health bureau, there would be a brand new practice building for the family doctors, facilities for minor surgery, radiology, community nursing, physiotherapy and chiropody, premises for a dentist, optician and chemist, a base for social workers serving the Tipton area, and a Citizen's Advice Bureau. For the health care consumer, it means everything together at one convenient location. But that was just the start of it.

Just 300 metres from the health park itself, a new shopping centre will be developed by a private sector interest encouraged by the transformation taking place in Tipton.

Investment is being made in sports and recreation facilities, reinforcing the 'healthy Tipton' initiative. For example, refurbishment of the baths, establishing waterborne recreation facilities, a new sports and community centre, upgrading of established parks and much more.

Visiting the doctor, picking up a prescription, getting advice on welfare benefits and doing the family shopping can all be done quite easily in future. Quite literally, the heart is being put back into the town.

More ideas are still coming forward in Tipton. Some, it is hoped, will qualify for start-up funding from National Lottery funds. Jogging and cycle trails along the canal have a potentially strong health contribution to make by encouraging people to become more physically active and by helping to reduce stress levels. A sporting academy based on National Lottery-funded all-weather athletics facilities to be used by a consortium of the borough's five athletic clubs, together with a major Lawn Tennis Association complex, is also envisaged. Using spare land on the site to produce food crops, which can then be marketed and distributed through a food co-operative, could boost attempts to promote healthy eating.

Lessons from Good Practice

What are the lessons about the best and most appropriate policies to be used under particular circumstances? Clearly, local conditions vary as well as local aspirations and expectations about what is required so that there can be no single blueprint for success. Public policy-makers, company executives and community leaders therefore tend to pragmatically develop their own community policy strategies. They may consciously or intuitively develop mechanisms that derive or borrow from the approaches mentioned above, but which also suit local conditions. Differences of emphasis in different local initiatives imply that policy-makers can select the most appropriate elements from different approaches. In practice, policies tend to reflect various strands of thinking and experience.

The lesson of Tipton's Neptune Health Park is that health can act as a major force for community regeneration, but the agencies and professionals involved must be willing to listen to local people. They must also be flexible and willing to lower their traditional boundaries in a spirit of genuine friendship. This experience seems to be consistent with that of the other partnerships mentioned in the introduction.

Within Sandwell, the Tipton experience has shown the benefit of adopting this approach. It may now be transplanted to the other five towns that go to make up the Metropolitan Borough. The lessons from this experience are about progress and working together. The excitement which projects such as those in Bolton and Tipton have generated springs more from the fact that all the key partners have been able to sit down with one another and discuss policy options to create a shared vision and pride in the community.

Finally, we recognise that this chapter provides only a snapshot of a situation that is constantly changing. The present government has initiated a wide-ranging series of policy changes and continues to produce policy proposals that will influence future initiatives. We have thus attempted here to provide an overview of issues that we feel will be of enduring interest during this period of change.

Key Issues and Actions

There are a number of important issues and actions arising from the above discussion:

- Initiatives in urban regeneration are implemented most successfully when programmes sensitively respond to local people including those with special needs and problems.
- The partnership model is an effective mechanism that ensures that practical polices can benefit whole communities. This is particularly the case when related policy areas need effectively to be co-ordinated and programmes efficiently managed.
- Community organisations play an important part in capacity building and encouraging the involvement and empowerment of people. This is especially so when policy-makers and programme managers have a clear idea of the factors that contribute to the development of a meaningful degree of community involvement in the process of urban regeneration.
- Local initiatives should spark a sense of purpose and pride in communities.

Notes

This paper does not necessarily represent the view of the former Tipton Challenge Partnership or Sandwell MBC.

Thanks go to Alec Morrison (Cobridge Community Renewal, Stoke on-Trent) who provided information about the European Network Eleven and to Susan Richardson of Bolton City Challenge for her local case study.

References

Benington, J. (1994) *Local Democracy and the European Union: The Impact of Europeanisation on Local Governance*, Commission for Local Democracy, London.

Bolton City Challenge (1995) *Progress through Partnership: Bolton's City Challenge Mid Year Review, 1995–1996*, Bolton City Challenge, Bolton.

Burns, D., Hambleton, R. and Hoggett, P. (1994) *The Politics of Decentralisation: Revitalising Local Democracy*, Macmillan, Basingstoke.

Business in the Community (BITC) (1986) *Business and the Inner Cities: How the Business Community Can Work With Others to Promote Better Opportunities in Our Inner Cities*, BITC, London.

Confederation of British Industry (CBI) (1988) *Initiatives beyond Charity: Report of the CBI Task Force on Business in Urban Regeneration*, CBI, London.

Department of the Environment, Transport and the Regions (DETR) (1997a) *Involving Communities in Urban and Rural Regeneration: A Guide for Practitioners*, DETR, London.

Department of the Environment, Transport and the Regions (DETR) (1997b) *Single Regeneration Budget Challenge Fund Round 4: Supplementary Guidance*, DETR, London.

Department of the Environment, Transport and the Regions (DETR) (1997c) *Regeneration Programmes: The Way Forward*, DETR, London.

Farnell, R., Lund, S., Furby, R., Lawless, P., Wishart, B. and Else, P. (1994) *Hope in the City? The Local Impact of the Church Urban Fund*, Centre for Regional Economic and Social Research, Sheffield Hallam University, Sheffield.

Geddes, M. (1995) *Poverty, Excluded Communities and Local Democracy*, Commission for Local Democracy, London.

Handy, C. (1997) *The Hungry Spirit: Beyond Capitalism: A Quest for Purpose in the Modern World*, Hutchinson, London.

Harcup, T. (1996) Looking after the health of the city, *International Cities Management*, September–October, p. 16.

Housing and Urban Development (HUD) (1995) US Department of (1995). Every community's a winner: Empowerment Zones and Enterprise Communities, *Community Connections*, March, pp. 6–7.

Imbroscio, D., Orr, M., Ross, T. and Stone, C. (1995). Baltimore and the human investment challenge, in F.W. Wagner, T.E. Joder and A.J. Mumphrey (eds), *Urban Revitalisation: Policies and Programs*, Sage, Thousand Oaks.

National Council for Voluntary Organisations (NCVO) (1996a) *The UK Voluntary Sector Statistical Almanac 1996*, NCVO, London.

National Council for Voluntary Organisations (NCVO) (1996b), *Annual Review 1995–96*, NCVO, London.

Taub, R.P. (1994) *Community Capitalism: The South Shore Bank's Strategy for Neighbourhood Revitalisation*, Harvard Business School Press, Boston, MA.

Tilly, C. (1974) *An Urban World*, Little Brown, Boston, MA.

Tipton Challenge Partnership (1997) *Annual Report, 1997*, Tipton Challenge Partnership, Sandwell.

Ware, A. (ed.) (1989) *Charities and Government*, Manchester University Press, Manchester and New York.

West, A. (1993) Putting Communities Streets Ahead, *BURA News*, no. 11, Winter 1993–94, pp. 18–19.

World Resources Institute (1996) *World Resources, 1996–97: A Guide to the Global Environment (The Urban Environment)*, Oxford University Press for The World Resources Institute, The United Nations Environment programme, The United Nations Development Programme and The World Bank, New York and Oxford.

7 Employment, Education and Training

Trevor Hart and Ian Johnston

Key Issues

Jobs for local people are the lifeblood if we want people to live in urban areas, and especially in inner urban areas. Similarly, the availability of suitable jobs is very high on most inner city residents' order of priorities. Moreover, it is now widely accepted that the competitiveness of a locality or area and its attractiveness to inward investors depend critically on its human resource. The basic core and vocational skills of the potential workforce and their attitude and motivation are critical. For this reason, education and training are key components of regeneration.

This chapter considers a number of major issues:

- the origins of urban employment problems;
- the development of urban labour markets;
- the evolution of urban labour market policy;
- the importance of urban labour market strategies;
- the implementation of policy;
- the likely future evolution of urban labour markets.

Urban Employment Problems

While, with commuting, it is possible to have economically successful capital investment in urban areas, without jobs for local people such areas tend to be sterile deserts at night.

Urban centres built on jobs linked with railway termini, docks and associated wholesalers and retailers have lost their purpose as the lorry and car have become the predominant means of transport. Mass production and ease of transportation away from congested urban centres mean that many jobs associated with provision of goods and services to the urban masses are now situated at the periphery or beyond.

Other cities have been particularly affected by technological change (for example, Sheffield and steel and Birmingham by the move from brass to plastic) and some by global competition (for example, Newcastle and ship-building, Bradford and wool).

Loss of Jobs and Unemployment

Much of the decline in traditional industries has been focused in urban areas. Problems, or perceived problems, likely to destroy jobs more quickly or make job creation in urban areas more difficult include:

- congestion which causes operational inefficiencies;
- high land values or polluted land so new building is expensive or difficult;
- public transport may be radial and may suffer from underinvestment;
- higher security costs;
- few skilled people;
- poorer basic education;
- perceived street crime and insecurity.

Unemployment may be higher, or more difficult to reduce, because:

- there are higher proportions of particularly disadvantaged long-term unemployed people;
- the impersonal nature of urban areas makes job matching more complex;
- hostels and other accommodation for disadvantaged groups are concentrated in the cities;
- concentrations of ethnic minorities may present language problems to potential employers;
- possibly a negative synergy brought about by a combination of such factors.

The job creation problems in urban areas are thus both distinctive and more acute than elsewhere. There is, however, often the political will present to invest in a solution. It does not particularly matter whether this is motivated by national or civic pride, by concerns about security, by a desire to utilise existing capital infrastructure or by a desire to produce equitable chances for the people concerned. All may lead to public authorities putting a higher value on jobs created in urban areas. Apart from buildings which are worthy of some national heritage status, if inner urban areas have served their usefulness to humanity, it is not clear from the standpoint of economic logic why they should necessarily be preserved.

Some successful urban programmes indeed do simply involve rehousing the jobless poor in thin dispersion in wealthier suburbs (for example, Chicago), or more directly training people to allow them to emigrate to other labour markets. But most programmes and policies involve attracting jobs into urban centres and equipping local people to compete for them through education, training, remotivation and job search skills.

Developments in Urban Labour Markets

The problems of urban labour markets are not of recent origin; they have changed in character over a period of many years, taking on different characteristics in periods of growth and decline, and manifesting themselves in different forms as a response to different phases of economic restructuring. However, there is some thread of consistency in policy responses. These are frequently prompted by a concern for the sharp social and economic inequalities that are evident in cities.

Table 7.1 **Population change for eight large cities**

City	Change for period (%)				
	1901–51	1951–61	1961–71	1971–81	1981–91
Birmingham	+ 49.1	+ 1.9	– 7.2	– 8.3	– 5.6
Glasgow	+ 24.9	– 2.9	– 13.8	– 22.0	– 14.6
Leeds	+ 19.3	+ 2.5	+ 3.6	– 4.6	– 3.8
Liverpool	+ 10.9	– 5.5	– 18.2	– 16.4	– 10.4
London	+ 25.9	– 2.2	– 6.8	– 9.9	– 4.5
Manchester	+ 8.3	– 5.9	– 17.9	– 17.5	– 8.8
Newcastle	+ 26.1	– 2.3	– 9.9	– 9.9	– 5.5
Sheffield	+ 23.0	+ 0.4	– 6.1	– 6.1	– 6.5
Great Britain	+ 32.1	+ 5.0	+ 5.3	+ 0.6	+ 0.02

Source: Champion and Townsend (1990), Stillwell and Leigh (1996).

Population Change

Table 7.1 shows the pattern of rapid growth and decline in the populations of Britain's largest urban centres. In addition to these changes in resident population should be added those who see these urban centres as their natural place of work but who live elsewhere: the commuters. Throughout the past 90 years, improvements in public transport and growing access to private transport have increased commuting flows into cities, partly through making it possible for former residents employed in the city to live elsewhere. This increase in mobility has added to the degree of social and economic polarisation in the cities, as those in the better rewarded segments of the labour market can exercise the choice to live elsewhere and travel into the city to work, while those in poorly paid or more marginal forms of employment have little choice but to live within the city.

The growth in population in the first half of the century resulted from changes in birth rates, improvements in health and changes in the economy, with the impetus provided by the continuing growth in some segments of manufacturing industry (for example, the motor industry) being complemented by growth in employment in the service sector. The depopulation of cities since the 1950s has been described by Champion and Townsend (1990, p. 161) as relating 'almost entirely to the forces of de-centralisation and counter-urbanisation'. In addition to increased commuting, a number of 'push' factors – such as pressures on land supply – play their part. There have also been, particularly in the 30 years up to the 1970s, planned dispersals of population from cities, at times reinforced by incentives and pressures to locate (or relocate) away from the main cities, and the development of substantial public housing estates on the periphery of cities. In more recent years, policies to encourage outward movements of jobs have been reversed.

The socially selective impact of commuting has in the past been exacerbated by immigration from the New Commonwealth, and of young people from other parts of the UK in search of work. On arrival, many of the new immigrants are relatively constrained in their choice of employment. While a certain amount of dispersal has occurred, the more marginal groups in the labour market concentrated into a number of deprived localities have suffered disproportionately from the decline in national economic performance reflected in rising unemployment figures from the 1970s onwards. In some urban cores, added pressure has been placed on certain groups by the gentrification of inner areas.

The Loss of Jobs

As Table 7.2 shows, the major loss of employment in cities between 1951 and 1981 has been in the manufacturing sector. This pattern continued in the 1980s with almost all the metropolitan counties and districts continuing to lose manufacturing jobs at a rate exceeding the national rate – a decline of 9.2 per cent between 1981 and 1991. In the 1960s much of this job loss

Table 7.2 **Employment change in major conurbations**

Period/sector	Overall conurbations		Great Britain
	000s	%	%
1951–61			
Total employment	+274	+3.4	+7.0
of which			
Manufacturing	–59	–1.7	+5.0
Services	+369	+7.7	+10.6
1961–71			
Total employment	–624	–8.3	+1.3
of which			
Manufacturing	–645	–17.2	–3.9
Services	–10	–0.2	+8.6
1971–81			
Total employment	–774	–11.	–2.7
of which			
Manufacturing	–927	–34.5	–24.5
Services	+89	+1.8	+1.8

Source: Champion and Townsend (1990).

can be traced to decentralisation, but in later years national trends and influences on manufacturing employment became much more important.

While there are many important general influences on the level of manufacturing employment, national economic policy exerted some significant pressures. In the face of falling domestic and export demand for manufactured goods, output and employment were reduced and many plants and businesses closed. These impacts were particularly felt in cities which were disproportionately dependent on manufacturing employment. The decline in manufacturing was exacerbated by the fact that cities were seen as less competitive and attractive locations for investment than other parts of the country. A wide range of factors contribute to a locality's competitiveness and its attractiveness to businesses. Particularly important are the constraints facing companies in cities: problems of unsatisfactory sites and limitations on relocation or expansion are identified as having a significant influence (Fothergill, Kitson and Monk, 1985). These are often more severe than the constraints facing companies elsewhere and have often been the decisive factor in the closure of inner city plants. There was both an urban–rural and a north–south dimension to this pattern of advantage and disadvantage, serving to emphasise the importance of viewing cities in their regional as well as national context.

Unemployment

Table 7.3 **Unemployment by place of residence**

(GB = 100)	1951	1961	1971	1981
Inner cities	133	136	144	151
Outer cities	81	82	88	101
Free-standing cities	95	107	112	115
Towns and rural areas	95	93	90	90

Source: Hasluck (1987).

Treating urban areas as a single category conceals the differences which exist both between and within urban areas. Some areas suffer multiple aspects of deprivation. The cumulative spiral of decline that can result leads to greater polarisation between cities but more particularly between localities within them. Unemployment is often taken as the principal indicator of the economic health of a locality and, as Table 7.3 shows, it is inner city areas which have suffered considerable unemployment. Results from the 1991 Census show that it is cities – and particularly the key urban centres – which continue to suffer a burden of unemployment greater than is experienced in most other parts of Britain. There are also significant differences between urban areas. In the 1980s areas that previously had not suffered high incidences of unemployment, such as the West Midlands, showed rapid rates of increase. Also in the 1980s a clear north–south divide was in evidence. Since this period, patterns of decline and recovery have been less clearly spatially differentiated: by the 1990s London experienced unemployment rates equal to some traditionally disadvantaged areas.

The 1980s saw worsening economic conditions in many urban areas, with higher proportions of long-term unemployed indicating an increasing degree of concentration of the most disadvantaged (Robson *et al.*, 1994). Almost inevitably this spiral of decline will be associated with falling job opportunities. As significantly, employment in these afflicted areas is likely to be concentrated in the those types of jobs which are of a lower quality than might be found elsewhere. This qualitative segmentation of the labour market has been characterised by Hutton (1995) as the 30:30:40 society: that is, the disadvantaged unemployed or non-employed; those in marginalised or insecure employment; and the fortunate 40 per cent who have (currently) some form of tenured permanent employment. It is also the case that certain groups are less well placed to avoid the impacts of recession including the young, the old, the unskilled and the unqualified, those with an illness or disability, those with a poor employment record, and ethnic minorities.

This latter issue is of particular concern for a number of reasons. First, the concentration of ethnic minority groups in certain localities throws a particularly sharp focus on their labour market problems: for example, 20 per cent of the population of Greater London is from ethnic minority groups, with the proportion rising to 40 per cent of the population in boroughs such as Brent and Newham. Second, some ethnic minority groups suffer disproportionately from unemployment: for example, unemployment rates of 45 per cent are observable in South Asian groups in Tower Hamlets. Finally, there is a likelihood of the problems for some of these groups becoming worse, as a result of the fact that the numbers economically active are forecast to continue to increase (by around 18 per cent in the ten years up to 2001), while the white economically active population remains fairly constant (Green and Owen, 1995)

Finally, in spite of the fact that women's labour market participation rates have continued to rise, gender segregation remains at the root of some profound labour market inequalities. Women are the main component of the growing part-time workforce and women's hourly pay rates are still only 65 per cent of those for men (MacEwan Scott, 1994). Seen in the context of household incomes, these factors play a significant part in defining deprivation in the broader sense in urban areas.

Urban Labour Market Policy

In earlier years, national policies for tackling the problems in cities, and particularly unemployment, were to a large extent contained within a growing range of regional policy instruments. In the immediate post-war years regional policy focused on the localities which had been designated as Special Areas in the 1930s. An urban policy began to emerge from the conduct of 'urban experiments' from the mid-1960s, prompted by a rise in unemployment, the recognition of persistent poverty despite the development of the welfare state, and the emergence of social problems centred on race. Lawless (1989) sees the most influential of these experiments as being the Inner Area Studies and the Community Development Projects, in that they had a focus of targeting and co-ordination, but most particularly because 'they helped transform attitudes to urban deprivation' (Lawless, 1989, p. 7). The Urban Programme was created in 1968, but its method of operation was much revised in 1977, the year which also saw the publication of the White Paper *Policy for the Inner Cities*, (HMSO, 1977) arguably Britain's only coherent attempt at developing a strategy for cities.

The period of the Thatcher governments coincided with a retreat from traditional regional policy and the emerging importance of a distinct urban policy. This change in priorities was accompanied by a change in philosophy. The 1983 review of regional assistance made it clear that the government's view was that the case for intervention was primarily social. It also promulgated the view that the problems on which policy was to focus were a result of supply-side weaknesses which undermined competitive efficiency and adaptability: so, the causes of concentrations of persistent unemployment could be traced to labour market rigidities, and to a lack of enterprise, reflected in low rates of new firm formation and technological innovation. From 1984 the focus of regional policy became much more concerned with the promotion of indigenous growth.

This change in philosophy was reflected in policies directly focused on the labour market. Two White Papers issued in 1985, *Employment: The Challenge for the Nation* (DoE, 1985a) and *Lifting the burden*, (DoE, 1985b) both stressed the theme of competitiveness and how it could be enhanced by the pursuit of neo-liberal policies. This approach was advanced in subsequent labour market strategy papers. For example, in a foreword to the 1992 White Paper *People, Jobs and Opportunity*, (Employment Department Group, 1992) the Secretary of State for Employment noted 'We must widen even further the choices and opportunities for people at work. We must stimulate enterprise and encourage the trend towards individual initiative in every aspect of working life.' In 1990 Training and Enterprise Councils were launched, charged with promoting the pursuit of the goal of a more skilled and competitive workforce: many Training and Enterprise Councils have operational areas including main urban labour markets. Training and Enterprise Councils also enshrined the concept of private sector leadership.

The position of TECs illustrated the dominant approaches and philosophy underpinning the Conservative government's policy for urban labour markets. In addition to the primacy given to private sector leadership, influence and ethos, there was a stress on partnership approaches at the strategy-to-project level. At the same time, there was still strong managerial control exercised by central government, seeking to ensure value for public money by employing a range of performance measures. The philosophy was one which saw the solution to urban labour market problems as being located in improved competitiveness, for individuals and businesses. The neo-liberal focus tended to undervalue other outputs from urban employment policy such as improved health or improved environment, and also undervalued the achievement of consensus with stakeholders other than private sector businesses – for example, trade unions and community groups (Marquand 1996).

Following the election of the Labour government in 1997 a series of reviews of relevant policy areas was put in train. The introduction of Regional Development Agencies from 1999 was the first substantial institutional change to emerge, but a number of changes in labour market policy will have relevance to urban labour markets. Echoing the terminology that had become common in European Union policy statements but maintaining an attachment to supply-side measures, a key focus of concern became the 'socially excluded', and developments in policy were seen as playing a role in 'tackling poverty and benefit dependency and contributing towards improved growth and productivity' (HM Treasury, 1998 p. 2).

Developments in the labour market, including changes in skills requirements, changes in the gender balance of employment and the changing impact of the tax and benefit system have led to what is described as a new approach being adopted. This is seen as having three key elements:

- the 'New Deal' programmes to aid progress from welfare to work, focusing on a number of groups including 18–24-year-olds, long-term unemployed aged over 25, lone parents, partners of the unemployed, and the disabled;
- the introduction of the national minimum wage and changes to the tax and benefit system, intended to 'make work pay';
- continuing investment in education and skills development in order to align the UK skills base with the needs of the modern economy.

In the 1990s there has been an increasing realisation that cities must make the most of their unique local assets and leadership. For example, cities may well emphasise their cultural industries (Bianchini and Parkinson, 1993; Kearns and Philo, 1993), the role of sport (Kitchen, 1993), the role of higher education (Robson *et al.*, 1995; Armstrong and Grove-White, 1994), the promotion of urban tourism (Law, 1993), and the importance of city marketing (Haider, 1992; Paddison, 1993). All this has led to some rethinking about the role of local leadership (Judd and Parkinson, 1990), to a growing criticism of the paradigm of property-led regeneration (Turok, 1992), and to a more critical, analytical approach to partnership (Peck and Tickell, 1994). There is a growing emphasis on bottom-up approaches but embedding locally determined approaches within an environment where resources are substantially controlled by the central state represents a substantial challenge for urban labour market policy.

Labour Market Strategies for Urban Areas

The problem of unemployment 'occupies a central role in the "urban crisis" because it is both a symptom of the processes which have undermined the urban economies and an immediate cause of poverty, poor housing and other aspects of social deprivation' (Hasluck, 1987, p. 2). This quotation emphasises that, while labour market issues are a central concern, they are closely interrelated with many other aspects, which in turn are addressed by various public policy measures, and by agencies within and outside the public sector.

The fact that there are a growing number of agencies and initiatives in play makes the need for co-ordination and strategy more pressing. Effective local action requires careful analysis as an input to its design. Monitoring and other approaches to ensure and to demonstrate that the most effective use is being made of public funds depend on having a well-established strategy and objectives against which to measure performance. Finally, there is the pragmatic reason for adopting a strategic approach that a growing number of EU and UK public sector programmes require a strategy as an input to applications for funding. A successful approach to strategy is likely to be one which combines the exogenous weight of national and European programmes with the endogenous benefits of organic, rooted, bottom-up initiatives, fused together in a genuine and inclusive partnership.

Defining Job Creation

At local level job creation is best defined as the net jobs added in that particular area over time. Local authorities or other local agencies probably do not worry about whether the jobs have come from neighbouring areas or whether they are 'new' jobs. Indeed one purpose of urban policy is to attract jobs from more prosperous or naturally attractive areas into less advantaged areas. Another key measure in the evaluation of urban policy is the reduction in unemployment.

In principle jobs come from increasing demand for labour by attracting or creating jobs and/or by improving the supply of labour (for example, more skilled and better motivated potential employees).

The demand side for labour can be enhanced by:

- attracting inward investment;
- growing existing businesses, especially firms in the 10–100 employee size range; this can be assisted by measures such as the provision of advice and technology transfer schemes;
- creating micro-businesses through encouraging self-employment;
- temporary job creation through publicly funded schemes;
- expanding the public sector;

- reducing labour costs (both wage and non wage) and various forms of labour market regulation to increase the employment intensity of growth.

The first two or three approaches are most likely to hold potential for local agencies. Temporary job creation measures are expensive, and the jobs are, after all, temporary. Expansion of the public sector can be potentially counter-productive if high resulting taxes and rates deter inward investment and make existing businesses less competitive. Reducing labour costs, especially where this involves reducing wages, risks losing social consensus. If taken too far, particularly in urban areas, such an approach might lead to heightened forms of social conflict.

The supply side can be enhanced by:

- providing information to make the labour and education and training markets work better;
- improving basic education including English as a second language;
- developing vocational skills;
- enhancing confidence, motivation and job search;
- changing unemployment or other benefit to increase incentives to work – for example, overcoming the benefits trap to provide higher in-work benefits; by reducing benefits; by making access to benefits conditional on job search or undertaking community work; or by changing other benefits such as child care so single parents are free to work.

Even in times of recession, the labour market in Britain is very active, with about 7 million job changes per annum. The net result is that if someone becomes unemployed the chances are better than even of finding another job within six months even in a recession. However, the longer people remain unemployed and the less educational and skill qualifications they have, the lower their chances of getting a job become. Because of this it is important that supply-side measures are targeted on those that really need help.

Dead-weight, Displacement and Substitution

Funding agencies may refer to the potential waste arising where help is not effectively targeted as dead-weight, displacement or substitution (DDS). In very broad terms these have the following meanings. Dead-weight means that the person would have found a job anyway without the special help. Displacement means that although the person helped got the job, they stopped someone else from getting it. Substitution (or 'crowding-out'), means that the firm taking on the person receiving special help, such as an employment subsidy, survives but only by putting another unsubsidised firm out of business.

This produces profound problems for national policy-makers seeking to help unemployed people. They know that to show any economic return to the taxpayer, their programmes have to be particularly effective. In evaluating their programmes they know that, after DDS have been deducted, often only 10 per cent of the jobs claimed for the programme are truly additional. This puts a high priority on high-volume, short-term, cheap interventions.

However, if programmes are evaluated in social rather than economic terms unemployed people can have their individual chances of getting a job very significantly improved. They can be promoted to the front of the jobs queue, even though this is achieved almost inevitably by pushing someone else further back.

Local agencies can also hope to push the displacement and substitution out of their local area. Sitting in Jarrow, it does not matter whether a job gained in Jarrow has meant one lost in Durham, Dresden or Japan. Deadweight, displacement and substitution are also why inward investment looks a good bet at whichever level the policy-maker sits.

Taken to extremes, buying jobs through incentives and subsidies can significantly disrupt the free market and that is why within the European Union there are strict competition rules outlawing continuing job subsidies. It is permissible to offer temporary subsidies attached to disadvantaged individuals, but not to offer long-term wage subsidies on the jobs themselves. In general most policy proposals from local agencies understate the 'cost per job' gained. The main value of cost-per-job evaluations is in ranking schemes.

In practical terms there are three approaches local agencies can take which minimise dead-weight, displacement and substitution, namely:

- Try to create jobs engaged in producing goods or providing services sold to people from outside the local area. This means avoiding traditional local services like hairdressing, sandwich bars and most house-building.
- Aiming for sectors with long-term growth trends. In conventional services this means, for example, leisure, education and health. In manufacturing it means focusing wherever possible at the high-tech, high-value added end of the market.
- Train for skills in demand. With unemployment around 2 million the job seeker to vacancy ratio exceeds 10:1. The job seekers include people with a wide range of skills. However, in principle, if vacancies can be identified in the local area which employers cannot fill and unemployed people are trained to fill them, dead-weight is zero, the unemployed people are happy, and the business community is pleased. In practice, there are numerous barriers to achieving this ideal.

Multiplier Effects

Having examined the difficulties it is worth also noting that success breeds success. If an inward investor establishes a company bringing 100 jobs there is the prospect of the wages of the new employees, together with the firm's expenditure on its supplies being spent to some extent on local goods and services, both sustaining further new 'linkage' jobs. That is providing, of course, that the company can be persuaded to use local suppliers and employ local people. The overall net economic effect is therefore the gross jobs created by the programme (or regeneration project) minus dead-weight, displacement and substitution plus associated multipliers (HM Treasury, 1995).

It is not really possible to estimate the multiplier effect since the range of effect varies from 0 to 100 per cent depending on the nature of the incoming investor's company and where it sources its supplies. Across a wide range of companies a rough average might be 30 to 40 per cent. However, the spending power of local employees has been estimated to add 6 per cent to the initial number of primary investor's jobs where the initial jobs were in metal goods (Owen, n.d.), and is perhaps more generally taken to be in the range 5 to 20 per cent.

Running Faster to Stand Still

As already mentioned, when additional jobs are created unemployment does not necessarily fall correspondingly. This is because as job prospects improve people at the margin may be encouraged to enter or re-enter the labour market, for example a spouse contemplating re-entering the labour market after raising a family, or a student making a choice between job search and further study. If a local agency is successful in creating local jobs, some or even many of the jobs may go to such job seekers, paradoxically leaving the local rate of unemployment unchanged.

The reverse can also happen if the 'feel good' factor is missing or if the economy is in recession. Unemployment can fall even though new jobs are not created, or it can fall faster than the rate of job creation. The latter happened in the British economy in 1994 when growth was well under way but the 'feel good' factor was missing. Employment rose by 266,000 but unemployment fell by 317,000, though this must be considered a very unusual combination.

Temporary and Part-time Jobs

Many projects create temporary jobs, for example in construction. These are clearly not as valuable as permanent jobs sustainable after the project has finished. But temporary jobs do bring wealth into the area and do offer long-term unemployed people a stepping stone back into the labour market. From the point of view of evaluation (HM Treasury, 1995), they require the introduction of the notion of job years. Part-time jobs have been a very significant source of job growth over the past decade. Many people with part-time jobs want to work part time, rather than full time.

The Informal Economy

A further complicating factor in urban economies lies in establishing the extent of the informal cash economy. As many as a third of those currently defined as unemployed may have some means of supplementing their unemployment benefit, especially where unemployment is high.

Certain studies have shown that many unemployed people have informal skills, such as painting 73 per cent, cleaning 59 per cent, gardening 44 per cent, car maintenance 30 per cent (Lawless, 1995). This presents opportunities for local agencies both to convert some informal activity into self-employment in the recognised economy, and to enhance the real wealth of poor neighbourhoods by deliberately raising skill levels in occupations useful for survival in the informal economy.

Other Aspects of Policy

In designing job creation projects it is important:

- to keep in mind the objectives of other partners;
- to tailor the skill requirements to the experience and skills of local residents (for example, low skill low demands, or accepting young people as their first step into the labour market);
- to aim for sustainability (for example, eventually self-financing or identified long-term public funding);
- to seek wide community support from local residents.

Sustainability is not easy to achieve. The evaluation of the Handsworth Task Force (PA Cambridge Economic Consultants, 1991) showed that only one in three projects lasted more than two years after the Task Force and that the job effect wasted rapidly. This illustrates the need for an effective exit strategy of which the most enduring is clearly likely to be commercial viability.

Initial Education and Training

In local labour markets where jobs are scarce it is important that pupils gain a sound basic secondary education particularly in mathematics and English. Inevitably it is in schools in just such labour markets that truancy tends to be high and academic attainment tends to be low. The value to pupils of working towards educational attainment may be less obvious where work opportunities are few.

The last decade has seen the introduction of quantitative benchmarks. Standard assessment tests and examination league tables can be used to match schools with comparators in similar locations. Of more strategic usefulness are the National Education and Training Targets which count all the qualifications gained by the entire cohort of young people including academic general (General National Vocational Qualification – GNVQ) and vocational qualifications (National Vocational Qualifications – NVQs). Any school, college or training provider can compare its current attainment against either the national or local target and use this information to set quantified goals for future improvement. The National Targets now include specific targets for core skills – numeracy, communication and use of IT – which employers say are important.

There is a vital role in regeneration for the business community to help the education world to break the vicious circle of 'failing schools'. Through active education-business partnerships the business community can contribute among other things to:

- strategies and analysis;
- improved management of schools;
- modern examples for the curriculum, modern materials and machines;
- continuing learning opportunities for teachers, particularly about the world of work;
- work experience for pupils and students;
- compacts to set goals and reward pupils;
- mentoring to support at-risk pupils/students.

One task is to convince pupils in deprived areas that there is some link between education and jobs, and that they too can succeed. They also need to be persuaded to stay on in education and training as long as possible to progress as far as their personal potential allows.

Post-Compulsory Education

Once they reach 16, young people face bewildering choices including:

- full-time or part-time study;
- education or training;
- academic qualifications, GNVQs or NVQs;
- in-school, college, private training provider or from an employer.

A full range of work-based apprenticeships and youth training offering full vocational qualifications should be available. These choices are important and many local careers services are properly equipped to help young people make them. A characteristic of urban deprivation is underuse of the advisory support that is available, and one aspect of a local employment policy might involve special outreach by the careers service and adult guidance services.

At a national or a local level the more young people can be persuaded to stay on post-16 in education or training the better – in the short term they are not competing for full-time jobs while they are studying, in the long term they are helping to raise the nation's skill base which will improve competitiveness and help gain jobs in the future.

Most urban areas have ready access to colleges of further education or universities and these can be an important source of continuing education and training for adults. For unemployed people these opportunities are often free and designed to fit within the tuition time limits imposed by unemployment benefit/job seekers allowance regulations so that benefit is unaffected (16 hours per week under the job seekers allowance for further education: full-time higher education is not allowed but career development loans are available to assist some mature students).

Technology Transfer

Education institutions also have a major role to play in helping local firms keep at the leading edge of technology and are large inner city employers in their own right.

Information and Analysis

A prerequisite for effective action is a clear understanding of problems and their underlying causes; this can then complement an understanding of which are the relevant agencies of influence, and their capability and capacity for action. It is also important that the progress of interventions should be tracked and this may require the establishment of a baseline of labour market conditions.

An important preliminary question is: which local labour market? Clearly the geographical extent of labour markets varies with skill and occupation, among other things; to some extent this point has already been established in the discussion of commuting earlier in this chapter. Whatever decision is reached, it is often also partly determined by the availability of data.

There are many sources of data on the demand and supply sides of the labour market, and there have been several unsuccessful attempts to develop and establish systems which integrate these data sources. This means that it is usually the responsibility of individual agencies to assess their information needs, and to gather together appropriate information from a diversity of sources.

In the majority of cases organisations will be reliant on secondary data sources in attempting to understand or monitor their local labour market: the availability of data by geographical area is set out in Table 7.4. When using this data it is important to be aware of its limitations. Data which is to be used for monitoring purposes needs to be at the very least consistent between time periods. There is also difficulty in using these secondary sources to describe a baseline picture, as they are not always as inclusive as might be desired, nor do they provide a fine grain of analysis.

The establishment of a network of TECs has added another source of data, although it has to be said that the majority of TEC labour market assessments are a mixture of primary and secondary sources. The data added by the TECs themselves tends to be 'soft' and their assessments may not meet the needs and priorities of all organisations.

In the absence of appropriate secondary data, primary research may be necessary. In the case of local demand data, it is frequently necessary to undertake employer surveys. On the supply side, there is often a need to supplement the regular statistics from the monthly claimant count, either by conducting a skills survey or a skills audit.

Table 7.4 **Regularly available local labour market data by geographical area**

Data/level	Establish-ment	Ward	Post code Sector	District	County	Travel to work area	Region
Employment:							
Annual employment survey	x	x	x	x	x	x	x
Labour force survey				some	x		x
Unemployment:							
Claimant count		x	x	x	x	x	x
Labour force survey				some	x		x
Unemployment unit		some	some	some	x	x	x
Vacancies					x	JCA	x
VAT registrations				x	x		x

Note: JCA = Job Centre Area

A final dimension of interest is a consideration of the future: in addition to attempting to tackle existing problems, organisations will be interested in considering likely developments in the local situation. Unfortunately, in many ways forecasting is more problematic than analysing or monitoring the local labour market. Typically, forecasts are related to or based on one of a number of national forecasts, informed by a variety of sources of local data, and disaggregated into demand and supply elements.

Key Actors and Agencies in the Local Labour Market

Beyond education and training providers, without doubt the three most significant players are local authorities, Training and Enterprise Councils, and the Employment Service.

On the business-led front government have accepted TECs as the key private sector agency, but of course in many places Chambers of Commerce, Chambers of Trade, traditional local associations (for example, in Sheffield the Cutler's Company) or single industries, (for example, British Coal Enterprise), can be very important partners. Over 90 per cent of TEC funding comes from national government although a tiny minority of TECs, most notably Northumberland, are developing with genuine private sector funding.

Other directors on TEC boards can be from local authorities, education, trade unions, voluntary bodies or other parts of the public sector. So in some ways TECs are a form of formalised community partnership focused specifically on skills, small firm formation and growth, and local economic development.

Training and Enterprise Councils are expected to:

- lead in local strategic policy and encourage local progress towards the National Education and Training Targets;
- be major partners in education business partnerships and to offer strategic guidance on Further Education provision;
- support local careers service arrangements;
- contract provision for work-based training and apprenticeships;
- contract provision for training and work experience for long-term unemployed people;
- stimulate lifetime learning for adults;
- encourage good company learning and training practice, and to act as assessors of the Investors in People training and development;
- encourage small firms and self-employment, and to be the strategic local lead contractor in the Business Link advisory network for small firms;
- encourage local economic development more generally.

Although most of TEC funding is for the specific purposes described above, TECs have considerable discretion to target their main programme imaginatively on particular localities and also to spend any surpluses in ways they wish. There has inevitably been debate between TECs and other partners, as well as with the government, about how much discretionary funding it would be prudent for TECs to spend each year.

From 1999, the context for local skills and training strategies will change, initially with the emergence of the Regional Development Agencies, and later through the implications of the White Paper *Learning to succeed* (DfEE, 1999). Exactly how such changes affect the development and working of local partnerships remains to be seen.

There are two basic tasks for the Employment Service. First, through the extensive network of Job Centres, the Employment Service pays benefit to unemployed people. Second, the Employment Service acts as a free public broker between the job seeker and employers with job vacancies. The Employment Service can provide information on trends, provide a placement service to potential inward investors and an advance guidance service to redundant employees.

Of more interest to local agencies pursuing local regeneration is the Employment Service role in helping each job seeker develop a Job Seeker's Agreement. This means every unemployed benefit claimant is interviewed regularly and that the Employment Services deploys opportunities for job search skills, work experience, work trials and subsidised probation, training and Further Education. The Employment Service can therefore act as a shop window of local opportunities for local people, as well as delivering national programmes locally.

Given their strategic overview and democratic legitimacy it is not surprising that local authorities lead or convene most partnerships and have deep interests in education and jobs. They are responsible for compulsory state education, they are often partners running Careers Services, they subsidise some adult Further Education and offer some adult students discretionary grants, and they administer grants and fees for Higher Education. Some Authorities run adult guidance services.

Local authorities are:

- landowners on a massive scale (for example, Birmingham City Council owns a quarter of the land area of Birmingham);
- planning authorities;
- capable of servicing land;
- developers of managed workspace.

As one might expect, therefore, local authorities and TECs form the heart of most local partnerships which have labour market objectives. Since job creation is at least a subsidiary objective for most regeneration activity, the potential for synergy is for real. For example, environmental improvements, housing maintenance, thermal insulation, care and mentoring all need labour. It does not matter that many of the jobs are temporary, because they give local people the opportunity to regain confidence, to break out of benefit dependency and to acquire a recent work history.

The presence of TECs in partnerships facilitates access to business advice and participation from TEC boards or wider business networks, and therefore provides exposure to new ideas, expertise, leadership and champions. However, the business orientation of TECs does not always fit easily with the public and community partners' interests and attitudes.

As just noted, multiple objectives are a strength. TEC directors from business have been uncomfortable in joining partnerships with objectives which have in some way watered down what the TEC might have wished to try and achieve if focused solely on its own mission and objectives (Environment Select Committee, 1995). Failure to win challenge bids, focusing on hard-to-place clients in output-related funding regimes, or even spending TEC reserves to the extent that the TEC reports a net 'loss', are all difficult for results-oriented business people.

While the presence of TECs may encourage genuine private sector resource contributions, it does not bring them automatically. Indeed in the labour market arca large-scale private contributions are unlikely, compared for example with the prospects of private asset formation through property development. None the less, some private sector companies may be prepared to back education or training projects, such as compacts, as part of their broader community relations strategies.

Funding

Although in principle the availability of funding and the conditions attached to it should never shape strategy, or projects responding to local needs, in practice the availability of funds often determines feasibility. Public sector funding for tackling problems of urban labour markets can come through a variety of sources – direct from central government, via quangos or from local government.

As important as the absolute amount of public funding may be, changes in the way funding regimes operate is of almost equal importance. The 1990s have seen the growth in the importance of a number of approaches, notably the importance of partnership, Challenge funding, the role of the private sector, and output-related funding for employment schemes. The partnership principle in urban regeneration has a long pedigree, particularly in relation to capital projects For example, the Private Finance Initiative launched in 1992 was expected to yield £14 billion for capital projects by 1998–99. Challenge funding has become particularly important in urban regeneration projects. The first bidding round for new regeneration projects supported by the Single Regeneration Budget (bringing together 20 previously separate government programmes from five departments) was held in 1994, when 200 bids were successful.

The partnership principle has also been given significant prominence in EU funding regimes. The partnership sought here is one involving the public, private and voluntary sectors. These partnerships are structured within a programme for the regeneration of the area, and the programmes themselves are meant to be the product of partnerships. European Union programmes typically provide no more than 50 per cent of the cost of projects, emphasising the importance of the availability of matching funding.

The Implementation of Strategy

As may be expected after around 30 years of developing urban policy, there is a rich variety of approaches to tackling the problems of urban labour markets. The case studies presented below are in no way scientifically representative of this rich variety, but do give an indication of what is possible.

Box 7.1   Managed Workspace: Bradford City Challenge

Traditional managed workspace, where both new and established small businesses can rent space and share common services such as reception, secretarial,
fax and phone answering, can have a significant local impact. Bradford City
Challenge's 'Commerce Court' offers space for 52 small business units and is
situated adjacent to a council estate with a long history of problems with
unemployment. Opened in June 1995 and managed by Bradford Chamber of
Commerce, the development achieved an occupancy level of 68 per cent within
six months. The cost per job brought into the area is estimated to be £7,300.

Of the many attempts to categorise urban policy interventions, that used
in *Action for Cities* (HMSO, 1988) is perhaps as useful as any. This grouped
the central government policy measures into five groups:

- Helping Businesses Succeed – concerned with encouraging enterprise
  and new business and helping existing businesses grow stronger. This
  thread has continued to assume great importance, in the work of TECs
  and others, and is a central task of Business Links. As well as using
  funding from the UK public sector, a number of EU programmes are of
  importance here. A range of agencies are involved from the public,
  private and voluntary sector, and it is perhaps the case that the
  educational/training perspective is seen as being of growing importance
  in the quest for competitiveness.
- Preparing for Work – focused on the aim of improving people's job
  prospects, motivation and skills. This strand was focused on the school
  leaver and identified school–industry links and established youth training among its main areas of action. While these types of activity remain
  important, there is a growing emphasis on the skills of those in work,
  with the objective of maintaining and developing appropriate skills to
  ensure continued employment and competitiveness in the labour
  market.
- Developing Cities – included the aims of making areas attractive to
  residents and businesses by tackling dereliction, bringing buildings into
  use and preparing sites and encouraging development. This action continues to be an essential component of many of the activities funded
  from the Single Regeneration Budget and from the European Social
  Fund (ESF). Overcoming the adverse impact on local competitiveness
  associated with dereliction continues to be an important part of job
  creation.
- Better Homes and Attractive Cities – concerned with improving the
  quality of housing and making inner city areas attractive places to
  live. It is an important adjunct to labour market policy to retain or
  attract those in higher skilled and better paid employment to inner
  city areas.

Box 7.2 Positive Action Consortium, Bristol

The project addresses the underperformance of ethnic minorities by placing them with companies. It uses £2,000 per trainee to lever an £8,000 contribution from companies. Eighty-eight per cent of clients had been unemployed for over two years and 49 per cent were from the Task Force area. The project had 1,500 trainees and 90 per cent got work at the end of their placement (but not necessarily with the host company). It had the involvement of 60 plus companies including Hewlett Packard and NatWest Life, all of whom were cold canvassed. It had funding from SRB, ESF and TEC core funding. Its aims: to be a one-stop shop for IT training, enterprise training and careers advice; to build some workshops; and to get more funds from the private sector (their funding is currently all time-limited, so they feel a bit insecure).

Integration of Policies

The development of a network of Government Offices for the Regions and the launch of the Single Regeneration Budget continued the move towards closer co-ordination and integration of government policy initiatives: Regional Development Agencies offer scope for closer integration at a regional level. The significance of European programmes and funding for urban labour markets has grown with a developing awareness of these programmes among public and voluntary agencies, and with the extension of eligibility to some urban areas such as parts of London. The ESF continues to be the most effective EU labour market policy instrument: in addition to supporting the mainstream UK programmes for youth and long-term unemployed, it often provides the means to fund innovatory actions focused on marginalised groups.

It is a continuing theme in the evaluation of UK and EU programmes that there is much to be gained by identifying good or best practice in urban regeneration, as well as by measuring success. Good practice should be seen as applying to both methods of operation as well as the programme content. However, models of good practice should only be applied in a way which respects local needs and circumstances.

Future Evolution of Urban Labour Markets

Urban labour markets are likely to follow national trends but also to show some distinctive characteristics in terms of particular growth sectors. In the national labour market there is a persistent trend away from unskilled and craft jobs towards professional, managerial and technician posts. While this will be reflected in inner urban areas also, it is exactly these levels of employee who can afford to commute from higher-cost housing in the suburbs or beyond.

There is also a trend towards a more flexible labour market, with large firms tending to contract out more work and for more jobs to be of a

---

Box 7.3   Wigan Borough Partnership

This partnership brings together the Metropolitan Borough Council, the local Chamber of Commerce, Metrotec, and others. They have thoroughly analysed their whole area in terms of social and economic indicators and the skills profile of their working population. Their focus has been on what brings wealth to Wigan. They found their current economy had the following structure:

Category	Employers	Employees
Wealth creation	15%	26%
Suppliers to wealth creators	17%	13%
Personal and domestic traded services	50%	31%
Publicly funded community services	17%	27%

Their strategy is to focus their efforts on improving the effectiveness of the existing wealth creating employers (for example by training and the Investors in People standard and by advice through Business Links), and by attracting inward investment. They have clear targets by 2005 to achieve the provision of wealth creating employment through:

- attracting 12,000 new jobs to the borough;
- creation/maintenance of 20,000 jobs among existing and newly formed businesses;
- education and training of borough residents.

They intend to measure their progress towards these targets.

---

part-time, temporary or short-term contract nature. The notion of all members of a family working part time indicates a priority for improvements in crèche and nursery provision.

Growth in employment is unlikely to come from large firms growing, since most are still downsizing and cost-cutting. The simple formula for corporate strategies – half the staff paid twice as much for producing three times as much – is set to continue. Net new jobs are generally thought to be likely to come from the growth of very small firms into medium-sized firms. Where these focus on high-tech, high-value added, high-knowledge activity they are most likely to add permanent employment and wealth creation in a global economy.

Net new jobs from self-employment and new micro-businesses are less likely but, as with training and other measures to help the unemployed, such opportunities may be a legitimate way of helping individuals into employment. Micro-businesses may also succeed in competing to win jobs in services from adjacent areas.

In terms of sectoral differentiation it is possible to think of urban labour markets being strong and becoming even stronger in:

- provision of Further and Higher Education, and associated with this a strong local service of technology transfer to companies through technopoles, science parks and the like;

- health care – being at the public transport hub helps;
- culture, entertainment, sport – again being at the public transport hub helps;
- heritage or theme retailing and tourism;
- the 24-hour city – supply of goods and services round the clock is a feature which may have further potential;
- local goods and services for local residents;
- more ambitiously, building on local strengths in sectoral terms to attract and grow new wealth-creating companies.

Two former strengths of urban areas, financial services and niche retailing, are both likely to suffer from the information revolution as more and more transactions can be made at a distance. However, the development of interest in environmental issues may hold advantages for cities which are alive to the possibilities presented by changing preferences and a developing legislative and policy framework.

Key Issues and Actions

- Population movements and economic changes are leading to cities becoming increasingly polarised, economically and socially.
- Cities have some unique assets as centres for service provision and consumption, and future development must maximise these advantages.
- Solutions to problems need to address issues of education and training as well as job creation.
- Local action must adapt to changed national labour market policies, which now emphasise supply-side measures instead of demand-side and favour partnership approaches over corporatism.
- The growing number of agencies increases the need for co-ordination of actions and interventions at the local level.
- Develop a clear understanding of trends in the local labour market and its inherent strengths and weaknesses.
- Map the pattern of actors and agencies influential in the labour market and the resources they bring with them.
- Work with others in the development of local labour market strategy as a basis for integrated local action, including the involvement of private and community sectors.
- Establish mechanisms and measures to evaluate the impact of interventions and initiatives.

References

Armstrong, H. and Grove-White, R. (1994) The economic and environmental impact of Lancaster University. Paper to ESRC Urban and Regional Economics Seminar Group, Leeds, September 1995.

Bianchini, F. and Parkinson, M. (1993) *Cultural Policy and Urban Regeneration: The Western European Experience*, Manchester University Press, Manchester.

Champion, A.G. and Townsend, A.R. (1990) *Contemporary Britain – a Geographical Perspective*, Edward Arnold, London.

DfEE (1999) *Learning to succeed*, DfEE, London.

Department of the Environment (DoE) (1985a) *Employment: The Challenge for the Nation*, Cmnd 9474, HMSO, London.

Department of the Environment (DoE) (1985b) *Lifting the Burden*, Cmnd 9571, HMSO, London.

Employment Department Group (1992) *People, Jobs and Opportunity*, Cmnd 1810, HMSO, London.

Environment Select Committee (1995) *First Report on the Single Reneration Budget*, HMSO, London.

Fothergill, S. Kitson, M. and Monk, S. (1985) *Urban Industrial Change: The Causes of Urban-Rural Contrast in Manufacturing Employment Trends*, HMSO, London.

Green, A.E. and Owen, D.W. (1995) Ethnic minority groups in regional and local labour markets in Britain: a review of data sources and associated issues, *Regional Studies*, Vol. 29, no. 8, pp. 729–36.

Haider, D. (1992) Place wars: new realities in the 1990s, *Economic Development Quarterly*, Vol. 6, no. 2, pp. 127–34.

Hasluck, C. (1987) *Urban Unemployment: Local Labour Markets and Employment Initiatives*, Longman, London.

HM Treasury (1995) *A Framework for the Evaluation of Regeneration Projects and Programmes*, HMSO, London.

HM Treasury (1998) *Pre-Budget-Report*, HMSO, London.

HMSO (1977) *Policy for the Inner Cities*, Cmnd 6845, HMSO, London.

HMSO (1988) *Action for Cities*, HMSO, London.

Hutton, W. (1995) *The State We're In*, Jonathan Cape, London.

Judd, D. and Parkinson, M. (1990) *Leadership and Urban Regeneration*, Sage, Newbury.

Kearns, G. and Philo, C. (1993) *The City as Cultural Capital, Past and Present*, Pergamon, Oxford.

Kitchen, T. (1993) *The Manchester Olympic Bid and Urban Regeneration*. Proceedings of the Town and Country Planning Summer School 34–38, RTPI, London.

Law, C.M. (1993) *Urban Tourism: Attracting Visitors to Large Cities*, Mansell, London.

Lawless, P. (1989) *Britain's Inner Cities*, Paul Chapman, London.

Lawless, P. (1995) Inner-city and suburban labour markets in a major English conurbation: process and policy implications, *Urban Studies*, Vol. 32, no. 7, pp. 1097–125.

MacEwan Scott, A. (ed.) (1994) *Gender Segregation and Social Change: Men and Women in Changing Labour Markets*, Oxford University Press, Oxford.

Marquand, J. (1996) *The Future for Tomorrow's People*, St William's Foundation, York.

Paddison, R. (1993) City Marketing, image reconstruction and urban regeneration, *Urban Studies*, Vol. 30, no. 2, pp. 339-350.

PA Cambridge Economic Consultants (1991) *An Evaluation of the Government's Inner City Task Force Initiative: Main Report*, DTI, London.

Peck, J. and Tickell, A. (1994) Too many partners . . . The future for regeneration partnerships, *Local Economy*, Vol. 9, no 3, pp. 339–50.

Owen, G. (nd) *The performance of local economic regeneration partnerships*, Unpublished paper.

Robson, B., Bradford, M., Deas, I., Hall, E., Harrison, E., Parkinson, M., Evans, R., Garside, P., and Robinson, F. (1994) (the Robson Report), *Assessing the Impact of Urban Policy*, HMSO, London.

Robson, B., Topham, F., Deas, I., and Twomey, J. (1995) *The Economic and Social Impact of Greater Manchester's Universities*, University of Manchester, Manchester.

Stillwell, J. and Leigh, C. (1996) Exploring the geographies of social polarisation in Leeds, in G. Haughton and C.C. Williams (eds) *Corporate City? Partnership, Participation and Partition in Urban Development in Leeds*, Avebury, Aldershot.

Turok, I. (1992) Property led urban regeneration: panacea or placebo? *Environment & Planning, b*, Vol. 24, no. 3, pp. 361–380.

8 Housing

Bill Edgar and John Taylor

Introduction

Housing is far more than somewhere to live. On the one hand, areas of monolithic housing, with inadequate amenities and few opportunities for economic activity, simply result in ghettos where those who manage to break the cycle of despair move away, leaving the remaining community poorer still. Many post-Second World War estates now provide classic examples of this spiral of decline. On the other hand, soulless commercial districts intimidate the ordinary citizen and regeneration without housing means areas without life, areas which become sterile outside normal working hours and which fall prey to vandalism, crime and the fear of crime, areas devoid of feeling and humanity rather than busy neighbourhoods buzzing with activity and a sense of community.

New housing can be a driver of urban regeneration, and decent housing is an essential ingredient of any regeneration scheme. Decent housing stimulates both physical and economic improvement, and the resulting enhancements in turn stimulate new investment and new opportunities as the urban environment once again becomes full of life and enterprise.

The importance of housing is illustrated by the fact that some 80 per cent of all development relates to housing and because where we live conditions so much of our daily lives (Gwilliam, 1997). Housing developments are, or should be, the trigger for the provision of facilities needed to meet daily requirements such as community, social, amenity, health care, and shopping. Clearly also, transport is needed for work and leisure pursuits. If we can provide good quality housing for a wide range of social needs, close to employment centres and other facilities, then we can help to regenerate our towns and cities and encourage a renaissance of urban living.

The key issues covered in this chapter are:

- the context for recent housing provision;
- how many homes are needed, where they might go and promoting urban living;
- links with health, crime prevention and education, urban design and social integration;
- finance, economic factors and housing construction;
- housing regeneration policy frameworks;
- implementation of strategies.

Context

Changes in the British housing market have been well documented elsewhere (McCrone and Stephens, 1995); the main issues that are of importance for urban regeneration are:

- housing stock – including questions of growth, and house condition;
- demographic change – including questions of household growth and household composition;
- labour market effects – including questions of migration patterns and housing demand patterns.

Housing Stock

The total number of dwellings in Britain in June 1995 was 22.2 million, an increase of 10 per cent since 1965. As well as changing in scale, the tenure structure of British housing has altered. With new construction typically running at between 90,000 and 110,000 per annum, the bulk of this change in tenure has been effected through house transfers from both the private rented and public rented sectors. During that period the balance of new dwelling provision has altered in favour of private sector provision (Figure 8.1); in 1979 the public sector accounted for 34 per cent of new dwellings while by 1995 this had declined to 5 per cent. Figure 8.1 also illustrates that the overall level of new housing provision has been declining since the 1960s by around 15 per cent (McCrone and Stephens, 1995; Wilcox, 1994).

During this period the condition of the British housing stock has improved. In 1965 almost 20 per cent of dwellings were below the tolerable or fitness standard (BTS) Below Tolerable Standard, by reason of a lack of standard amenities. By 1995 that figure had fallen to less than 5 per cent and the main reason for failure is more likely to relate to dampness rather than to a lack of standard amenities. While over a third of all BTS dwellings are in the public sector, 40 per cent are owner-occupied and a further 23 per cent are in the private rented sector (despite the fact that this sector accounts for only 6 per cent of total stock).

Since the mid-1970s the majority of capital expenditure on housing in both the public and the private sectors has been on the improvement and modernisation of the existing housing stock. Local authority capital expenditure has declined in real terms since 1979, resulting in a reduction in capital investment of about 61 per cent in England (Wilcox, 1994) and a reduction in Scotland of over 15 per cent on Housing Revenue Account (HRA) property and 33 per cent on non-HRA investment (Watt and Summers, 1995). Local authorities in Scotland have concentrated expenditure on modernisation of property which they wish to retain – increasing from 21,000 to almost 100,000 dwellings modernised between 1979 and 1994.

Figure 8.1 **New Dwelling Completions**

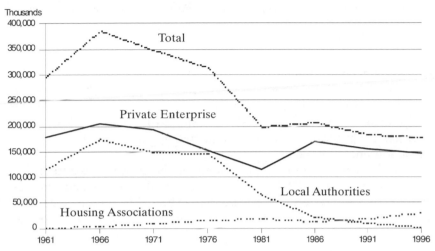

United Kingdom

Source: Department of the Environment, 1998.

Private sector investment in housing, in Scotland, was estimated at around £2.3 billion in 1993, with one-third accounted for by the provision of new housing and the remainder by expenditure on repair, maintenance and improvement.

In England and Wales, the Housing Investment Programme (HIP) is designed to create a comprehensive package of measures for local authority council housing stock. Evidence shows (Robson, 1994) that HIP allocations were cut in every year of the 1980s except for 1982/84, and that while expenditure rose over the period both allocations and expenditure fell in real terms. This pattern reflects an increasing reliance on capital receipts rather than HIP borrowing limits. Robson (1994) shows that the 57 Urban Priority Areas did not experience as large cuts as the rest of the country. Nevertheless they still lost almost half of their 1978/79 resources. Examining the relative shares of each year's HIP allocations shows an increased targeting of the UPA authorities whose share of the allocation rose from 46 per cent to 58 per cent between 1978 and 1989. In addition, around 75 per cent of the resources under the Estate Action programme have been allocated to the UPAs.

These changes to Britain's housing stock may be summarised in the following manner. Prior to the late 1960s new housing provision was at historically high levels and public sector provision was a high proportion of total provision. Poor condition and deterioration of the housing stock, by contrast, were also running at high levels and tended to be concentrated in the private (rented) sector. Policy responses were concentrated on increasing provision and redevelopment. Since then, new provision has been relatively low and mainly private in character, whereas house condition has been relatively good with the main areas of deterioration occurring in the maturing public sector. Despite this a high proportion of the housing stock is aged, while clearance and new provision is below 'replacement levels'. Policy responses have been characterised by concerns with tenure, affordability and access.

Demographic Change

Demographers have begun to refer to the period since 1965 as the 'second demographic transition', characterised by low fertility rates below replacement level, a corresponding increase in the proportion of the elderly in the population and an increase in smaller and single person households (Jones, 1990). These changes have been accompanied by changes in the nature and rate of household formation.

Britain currently has around 20 million households, which represents a 13 per cent increase over 1981. This upward trend is likely to continue, increasing by as much as 7 per cent by the year 2001. This rise in household numbers can largely be accounted for by the increase in single person households caused by demographic, cultural and economic factors (King, 1993).

Labour Market Effects

Labour market and demographic factors interact with housing market factors in influencing the nature and rate of deterioration of residential areas and the demand for new house-building. Some residential neighbourhoods may become relatively poorer over time as a result of a combination of ageing *in situ* (increased proportion of retired households), selective migration (out-migration of employed households) and labour market impacts (increasing levels of unemployment and low or unstable income). In areas of social renting, allocative mechanisms tend to accentuate these processes by allocating houses on the basis of housing need and thus concentrating the social and economic impacts of residential change (in residualised estates). Policies of community ownership, voluntary transfer and tenure diversification have been inevitable, if ideologically driven, responses to such problems.

These changes in household composition and formation are predicted by some (Lever, 1994) to indicate a likely shift in demand towards smaller dwellings and away from suburban villa lifestyle. In terms of locational choice it is predicted that preferences will change among certain sections of the population to favour urban living with access to the facilities offered by city centres. This coincides with a renewed interest in retaining residential populations in city centres.

Housing Needs and Urban Living

All of the above factors are crucial issues in determining the scale and composition of housing need, the level of demand for sites and the accommodation of the future population. Recent debate concerning the Department of the Environment's households projections for England and Wales (Vann, 1996) reflects a fundamental shift in attitudes towards house-building needs and the location of housing. Since the mid-1960s changes in the nature and rate of household formation have meant that the headship rate has increasingly become a much less reliable tool for predicting house-building needs. Analysis carried out by the Building Research Establishment (King, 1993) describes the pattern of net household increases in England and Wales from 1961 to 2011. Up to 1983 the pattern is one of steady growth, averaging around 130,000 households per annum. During the mid-1980s net household growth increased to around 200,000 per annum, fuelling the first-time buyers' boom of that period. Thereafter it is projected to fall back to more typical post-war levels, of around 100,000, for the remainder of the period to 2011. Recent revisions to household projections have been necessary because they have proved to be a poor indicator of subsequent house-building rates. A likely explanation is that the headship rates on which they were based reflected the exceptional headship rate boom of the mid to late 1980s. It seems likely that the fundamentally changed economic situation of the 1990s, in combination with this inaccurate assessment of headship rates, means that recent projections of households could be a significant overestimate. Significantly, in this context, although the late 1990s shows a falling away from the peak of the late 1980s, the overall pattern returns to the levels of household growth experienced in the 1960s and 1970s.

Three issues are considered in the remainder of this section:

- housing needs;
- the availability of sites;
- urban living.

Housing Needs

We have seen that housing is fundamental to regeneration, but how much of it is needed and where is it to go? There are at present around 22 million dwellings in the UK, many of which are substandard or in need of major refurbishment. Notwithstanding this, well-aired government research forecasts that 4.4 million new households will be required over the 25-year period to 2015. These projections are the result of demographic and social trends: more young people leaving home early, more divorces and separations and more old people surviving their partners for more years. You would have to be particularly Canute-like to believe that you could stop any of these things (Hall, 1998). The government has revised the figures several times in the 1990s, always upwards. If the figures are wrong, they are likely to be an underestimate (Best, 1997).

If additional houses are not provided, it is assumed that youngsters will be forced to remain longer in the parental home; couples will have to wait longer before embarking on the housing ladder and that the elderly will have to move in with caring sons and daughters. But this simplistic approach grossly underestimates the implications of the requirement. The essential problem is that, without adequate housing provision, the price of decent houses will rise and those without the means to purchase them will be forced back into substandard accommodation and sink estates.

Availability of Sites

Where will all the houses go? Monitoring in London has revealed that over a five-year period, as much as 50 per cent of house-building took place on 'windfall sites' not anticipated in development plans. This goes to the heart of the debate about brownfield versus greenfield development, and the lessons can be applied to all urban areas. Furthermore, some land will continue to be released by utilities and transport infrastructure, the rationalisation of sites used for schools and education institutions, hospitals, municipal buildings and, on a smaller scale, banking halls, post offices and similar quality buildings which can be converted to housing and other uses, creating diversity within the local area. From just one sector of the commercial market in London, it has been calculated that over 50,000 new dwellings could be created from redundant office buildings alone (London Planning Advisory Committee, 1998).

The biggest source of housing sites within the urban area will still be surplus industrial land, both empty premises and the 'classic' derelict sites, but the point is that these derelict sites are by no means the sole source and indeed a focus on these alone will seriously understate the amount of land available for recycling.

Urban Living

Cities are more than buildings and places where people simply survive. They are cradles of social and economic activity, where the very diversity of interactions creates new initiatives, new ideas and new energy. Cities have to be re-created as attractive places where those people with choice will want to live and work and where they will enjoy leisure and cultural pursuits (Taylor, 1997).

A much clearer vision of the future now informs the thinking behind many of the proposals for urban regeneration, and major cities have recognised the need to move away from their familiar industrial past to focus on service industries and more targeted, high-value production. Cities in this view of the future are becoming the locations where people gather and meet for entertainment, where information is assembled and exchanged, and where inventions and innovations are exploited as a basis for high value-added goods and services.

The exodus of people from town to country over the later part of the twentieth century has been one of the most important trends underlying the debate about the future of our cities. While the loss to our urban areas has been in the order of 90,000 people each year for the 15 years since 1982 (Champion *et al.*, 1998), the pattern is by no means uniform across the major metropolitan areas. In any case, this net figure merely represents the balance of much larger numbers moving in both directions. The urban exodus is not inevitable; 20 per cent changes in in- and out-migration would bring the exchange of people between town and country into balance (Reynolds, 1998). In seeking to explain why people move, the perceived attractions of the rural idyll are particularly strong, while many of the stated disadvantages of urban living are inconsistent and exaggerated. Indeed, whilst accepting that the most deprived areas are clearly in need of major regeneration to restore quality of life and confidence, for many groups of people, the advantages of urban living are very great. We need therefore to build on the growing mood of optimism and make particular provision in our towns and cities for the elderly, single people and young professionals.

Other Key Factors

In this section a number of other key factors and issues are discussed. In particular, emphasis is placed on health, crime and education; urban design; and social integration and community participation.

Health, Crime and Education

There is more than a century of solid evidence linking poor housing with ill health. Concerns about the effects of insanitary and overcrowded housing led to public health legislation in Victorian Britain, slum clearance programmes and the development of social housing. Still, despite very considerable public and private investment, over 1.7 million homes in Britain are officially deemed unfit. Even more are seriously lacking in proper maintenance and repair. All this must be tackled as well as the backlog in meeting current demand for social housing, before meaningful inroads can be made into the new household projections.

Crime presents problems and generates anxiety across all housing tenures and throughout the country. It is particularly severe on some social housing estates where the vulnerable and the disaffected come together. Noise, harassment, vandalism, theft and drugs are often cited as disincentives to urban living, although the relatively affluent and more rural areas are not immune.

Clearly, the causes of crime must be tackled through social and educational programmes, but the effects must also be addressed to create greater feelings of security. Most of the crime which is most troublesome takes place in the vicinity of the home. Schemes to combat anti-social behaviour, such as neighbourhood watch schemes and community safety strategies, have a part to play, while improved physical security and improved surveillance act as deterrents. In terms of physical regeneration, the extra cost of Secured by Design measures is very modest in new residential areas and refurbished estates and this can reduce very considerably the opportunities for crime. Overall, improved housing management and meaningful community involvement are often the key to tackling crime.

Good education lays the foundations for the personal skills and enterprise that are so fundamental for the well-being of the economy as a whole, and it is also critically important in terms of promoting socially responsible behaviour. Good schools are often the determining factor for many families, influencing the choice of housing locations for those moving into an area. However, whatever the education provision, poor and cramped housing conditions place many children at a disadvantage in reaping the benefits of schooling.

Urban Design

Commercial districts can be intimidating to the ordinary citizen. Large corporate buildings seek to outbid each other rather than welcoming mere human beings; and often these districts are segregated and intersected by urban expressways, creating both physical and psychological barriers.

The whole history of the city is living cheek by jowl, of densities which made for communities and the exciting and exacting use of space (Gummer, 1997). The most successful places for living and working are those which are compact, bringing together homes and work, making good use of infrastructure, and with the ability to adapt to changing fortunes without complete redevelopment.

Such projects will only work well if executed to the highest standards of urban design and with long-term management. Many mixed-use schemes have been of poor quality and have paid insufficient attention to the ways in which successful integration can be achieved. On the other hand, the best examples of comprehensive, mixed-use development have achieved environments that are undoubtedly more effective, valuable and sustainable than a piecemeal approach. Local authorities can assist with the drafting of suitable design guidance which, while avoiding unnecessary prescription, can articulate reasonable requirements for developers in terms of scale, the treatment of the public realm, connectivity, movement and related factors. Guidelines can be incorporated into planning and development briefs and should be the product of discussion with and the involvement of local interests.

Planning and Land Use

Much has been written in recent years about brownfield versus greenfield development. In many ways the essential features of sustainable communities are much the same in either scenario and the argument turns on wider environmental issues, such as the cost of new infrastructure, transport and the availability of the support facilities needed. It must make sense to plan for mixed developments and accommodate as much as possible of the projected growth within the existing urban framework, rather than create new settlements. This does not mean urban cramming, or high-density estates without the fundamental blend of support facilities; rather it means good urban design, with attractive public open spaces, good amenities, and making the best use of redundant buildings, and derelict and vacant sites. The presumption must be in favour of urban regeneration rather than out of town development.

There is a need to get away from sterile arguments about densification and focus on the interplay between better living environments, and sustainable patterns of movement. Explore new approaches to living which move from car-orientated suburban stereotypes to forms which are compact and attractive, reflecting the diversity of residents in cosmopolitan areas. Many sites in towns are within walking distance of central facilities and public transport, allowing a more urban and higher-density dwelling mix and significantly reduced car parking provision. The outcome will be an all-round gain by increasing the development values of such sites, so promoting their use; creating new types of housing attractive to small households, adding vitality and viability to town centres, and reducing car pollution while stimulating environmentally friendly transport. Such policies would increase housing provision by 50,000 homes in the Greater London area alone (Llewelyn Davies, 1997).

The planning system is still very mechanistic, yet planning can be a powerful force for regeneration if the tools available are used to enable high-quality mixed-use developments and sustainable forms of living. Too often, local plans are dominated by outdated notions of land use zones, urban density, car parking, highway standards. Stronger regional planning has a crucial role to play in tying opportunities for development into integrated transport and economic planning. Walkable neighbourhoods need to be well served by public transport in ways that can concentrate development along and around rail and bus corridors rather than new roads. It is equally important to target economic development activity at and around residential settlements in ways which minimise the need for motor cars.

Social Integration

The term social exclusion is relatively new to the housing debate in the UK. However, it is widely accepted that housing circumstances relate to, and contribute to, problems of social disadvantage generally. Poor housing contributes to the difficulties facing households and affects social integration Rather than assisting people to participate fully in society and increasing their chances of access to opportunities, housing and housing policies are increasingly regarded as contributing to the process of disadvantage (Lee and Murie, 1997). However, the links between housing tenure and disadvantage do vary across the country. The most deprived areas are not exclusively areas of council housing and relatively large proportions of ethnic minority groups live in the private rented sector.

Most cities have become predominantly a housing market for the poor in social housing, low-value owner-occupied, and privately rented accommodation. The introduction of middle- and upper-income housing into such areas could bring benefits that reinforce the momentum of regeneration. Apart from the economic impact, these new residents will have greater political awareness and good networks that connect them to those influencing local policies. Most housing schemes now aspire to mixed tenure and many local authorities make a proportion of affordable housing a condition of new planning consent for private sector developments.

Ownership and Community Participation

The lessons of the past must not be forgotten. Whether we have built up or out, much of the housing development since 1945 is now considered ripe for demolition while large numbers of terraced houses which escaped the bulldozers are now sought after and, with improvement, will have many years of life left in them. It was thought that citizens in urban areas would want to move away from city centres to huge complexes built on greenfield sites. It was to be a brave new world when new kitchens and new bathrooms would themselves make people happy. So acres and acres of high-rise flats were built, with nowhere to go and no one to know (Gummer, 1997).

It is now accepted that a fundamental precondition of sustainability in regeneration is to establish closer community involvement (Fordham 1995). The local authority planning process needs to be organised as a genuinely corporate process, one which emerges from a wide-ranging dialogue about needs and wants, long before any consultation draft is produced. Local communities need to be involved at every stage rather than as afterthoughts.

Finance and Housing Construction

Two key factors that exert considerable influence over the availability of housing are finance and the performance of the construction industry.

Finance

Housing investment in the UK is too low to satisfy the needs generated by the current backlog, additional household formation and the replacement of dwellings beyond economic repair. There is a vicious circle of low investment and high prices which is eroding post-war gains in housing provision, leading to severe housing shortages, and it is the lower income households which bear the consequences.

Housing investment during the 1980s and 1990s has fluctuated in line with the financial conditions as a whole. Prior to the early 1970s, housing investment was not directly related to the national fortunes and actually helped to stabilise the economy. Without sustained investment and the consequently reduced construction workforce, house-building during the boom periods is crowded out by commercial building activity. It is frequently believed that housing investment diverts resources away from growth-creating investment, but historical evidence in industrialised countries suggests otherwise: in the medium term, housing investment may actually increase national income through higher and more sustained employment within the sector. Housing investment therefore contributes to economic growth and stability (Ball, 1996).

The introduction of private finance into housing associations since the late 1980s has been most successful, and commercial interest rates at which funds are being invested suggest that associations are as good a risk as many large well-diversified manufacturing firms. However, the longer-term position could be different. Many associations assume that rents will rise faster than inflation, which may well be optimistic in some parts of the country. More significantly, in terms of the security of rental income, it is unlikely that the government will be willing to underwrite these increases and bad debt could therefore become more of a problem. Furthermore, a greater proportion of available housing association resources will have to be allocated to repairs in order to maintain standards and the asset value of the housing. The longer-term viability of private investment in social housing is therefore fundamentally dependent on wider government fiscal policy and the regulatory framework.

Raising private finance for local authority estates has not been a problem for housing associations, but valuations of properties on these estates tend to fall below the costs of development to a greater extent than elsewhere in the social housing market. This means that associations have to use reserves to maintain standards and to keep rents down. Associations' involvement therefore needs to be tied in to wider regeneration initiatives, in order to buttress their own investment. Conversely, associations often do not have a local management presence, in the sense of staffed area offices, and some local authorities are critical of associations' lack of involvement in wider issues affecting the estate. Partnership arrangements should therefore be structured to secure the long-term management of estates and thus also help to maintain the value of the asset base needed to provide the collateral required for continued support of financial institutions (Crook, Disson and Dark, 1996).

Housing and the Construction Industry

Historically, housing investment in the UK has been cyclical in nature. Boom periods have been related to technological and general economic circumstances. The major swings in demand experienced by the construction industry reduce its ability to build the required number of houses, to innovate and to reduce costs. Supply factors probably explain why house-builders respond slowly to upturns in demand and help to explain why prices rise so sharply during market booms. A greater degree of certainty in the application of planning policy coupled with government measures to promote long-term investment in housing will help to provide the stable conditions necessary for the housing industry to respond efficiently to the scale of the projected needs.

New Housing

Most new housing looks to the past rather than the future. The detached and semi-detached house are still the standard products of the volume house-builders, but these do not cater for the needs of the single person family, or older people living on their own. The problem in the UK is not so much an explosion of population, for overall this is modest, rather the much larger number of smaller households required to reflect demographic and social trends, and these are well suited to mixed-use developments.

There are a number of barriers to innovation, but more radical approaches to housing design have been explored through demonstration projects, such as the Joseph Rowntree CASPAR project, City Apartments for Single People at Affordable Rents, and in the design criteria set for the new Millennium Village on the Greenwich Peninsula. It is clear that housing construction can be rationalised to reduce costs while maintaining quality and individuality, and designs can be adopted which minimise the impact on the environment and produce more flexible interior layouts which can be adapted over time, without major structural alterations, to provide a home for life rather than one step on the housing ladder.

The private sector has a crucial role in adapting to these changes. It will be the innovation and imagination of the entrepreneur that will find the way to use brownfield land most effectively. Once they know that there will not be an easy way out – that release of greenfield sites is a thing of the past – then they will concentrate their formidable skills in discovering the most profitable ways of reusing land (Gummer, 1997).

The private sector has proved very successful in working in partnership with local authorities and housing associations in bringing forward very large housing and regeneration projects in very difficult areas. In the

Thames Gateway at Barking, 5,000 homes, with employment space, local training opportunities, and school and community facilities are being built by a major urban house-builder on a site with a high degree of contamination in places. Here much of the investment and the risk is taken up by the private sector. At Hulme in Manchester, the private sector has also played a leading role in this widely acclaimed housing-led regeneration project. The lesson is that, given the right conditions, and genuine working partnerships, which local authorities can greatly facilitate, the private sector is perfectly capable of delivering innovative, high-quality housing solutions which benefit from recent experience of mixed-use and mixed-tenure developments, and at costs to the exchequer much less than that of many less successful schemes in the past.

Rehabilitation Policy

The assumption underlying this analysis accepts the argument that 'the conceptual, evaluative and organisational frameworks for "doing-up" houses were not adequate to the task of reshaping areas, nor were "remodelling" frameworks appropriate for the strategic questions now confronting urban policy' (McGregor and MacLennan, 1992, p. 4). This implies that the strategic framework for housing investment in urban renewal has not been well developed in the past. It further suggests that there has been limited effective evaluation of previous housing renewal initiatives.

The evolution of housing rehabilitation strategy (a narrower concept than urban regeneration) is now well documented (Thomas, 1986). This shows that while policies moved from a focus on redevelopment to rehabilitation and then to regeneration, distinct phases of policy development can be identified from housing provision and upgrading of dwellings, to neighbourhood regeneration and thence to community regeneration. Broadly speaking, the concerns of urban policy reflected in each of these phases have evolved from: housing shortage (1960s), dwelling conditions (1970s), neighbourhood regeneration (1980s), and tenure diversification and community regeneration (1990s).

The changing emphasis to housing renewal can best be described by reference to the 1974 and the 1988 Housing Acts. These legislative expressions of policy indicate both the main approaches to housing regeneration which have been pursued and the changing emphasis of policy in relation to housing regeneration. The main approaches can be described as the use of administrative mechanisms, the area-based focus and the partnership approach:

- administrative focus: increasing the powers and funding of the Housing Corporation; creation of new bodies such as Scottish Homes; use of voluntary housing associations as the vehicle for housing renewal rather than local authorities;

- area-based focus: use of area-based initiatives initially in the inner cities and latterly in peripheral and inner city local authority estates;
- partnership focus: the increasing emphasis on private finance.

The changing emphasis of policy during this period can be described as:
- rehabilitation: inner city (private rented) housing (after the 1974 Act);
- modernisation: of large municipal housing estates (after the 1988 Act);
- tenure: diversification, creation of market-orientated independent rented sector and transfer of ownership of municipal housing estates.

It is evident from this overview that housing rehabilitation initiatives have changed since the 1974 Housing Act. The introduction of Housing Action Areas (HAAs), together with the increased role for housing associations, led to relatively small-scale single-purpose initiatives targeted at the rehabilitation of the areas of worst 'housing stress' with a five-year time-scale for implementation. During the 1980s larger-scale multifunctional approaches were attempted which recognised the need for longer time horizons. The partnership initiatives introduced by 'New Life for Urban Scotland' exemplify this approach. Thus while older housing regeneration initiatives had single-purpose objectives, short planning horizons and were not set within the strategic planning context of the urban system, later initiatives have tended to be multifunctional, larger, with longer time horizons and to recognise the ramifications for the wider urban system in which they are set.

Programmes during the 1990s have therefore tended to be based more on partnership, to involve community-based associations and co-operatives, to involve a mix of renovation and new build, to involve tenure diversification as an explicit strategy and to be undertaken over a longer time-scale than hitherto. It is axiomatic that such investment has involved a large and increasing level of private finance. However, these programmes have also involved a competitive bidding element using top-sliced funds (at least in England and Wales). In this context the impact of housing regeneration on less attractive residential areas becomes more of an issue than hitherto. McGregor and MacLennan (1992, p. 13) asserts that for initiatives developing in the late 1980s, especially in run down social housing areas, 'it has become increasingly important to assess an area's residential competitiveness in the context of likely demographic change'.

Framework for Housing Regeneration

Changes in housing renewal policy during this time have tended to reflect changes in the decision-making environment which can be summarised as:

- centralisation: changing role of local government;
- privatisation: private finance and non-governmental agencies;

- consumer sovereignty: growth of home ownership.

These trends can be identified in key elements to the approach to housing regeneration, for example:

- emphasis on area-based and local initiatives;
- labour market interaction of housing investment;
- partnership between local government and key agencies;
- growing involvement of private investment;
- community involvement.

It is arguable that our understanding of the factors that influence housing obsolescence has not been underpinned by rigorous empirical research during this period. Emphasis has been given to the physical factors of obsolescence and to the economic predicates. Much less attention has been given to social factors and to housing market factors such as tenure shift and residential or neighbourhood restructuring. While our knowledge of individual factors affecting housing obsolescence is better developed in some areas than in others, our understanding of the interaction of these factors is still in its infancy.

It is also arguable that the lack of adequate monitoring of initiatives, at least until very recently, has left an absence of information for the purposes of research and evaluation. Thus the ability to predict trends and to develop strategic plans for housing regeneration continues to be limited.

Despite this lack of understanding, the changes in the decision-making environment described above have resulted in significant changes in perception in relation to important issues affecting housing renewal. Three examples may be cited.

First, the growth of home ownership and consumer aspirations has led to change in the socially acceptable standards of dwelling condition and quality, even if these have not yet led to changes in the statutory definition of minimum dwelling standards. Such changes have been to include factors which affect the quality of life (such as central heating and double glazing).

Second, the shift in the role of local authorities from landlords to enablers has been reflected in a growth of public expenditure on housing maintenance and renovation. Although local authorities remain landlords for existing homes, the restriction on their ability to build new homes together with pressures such as tenants' rights and performance measures has led them to focus their efforts on maintenance of their remaining stock. Thus issues of dwelling maintenance become more significant in terms of public expenditure rather than issues of housing rehabilitation.

Third, the growth of private sector investment in housing association provision has led to the growing importance of life-cycle costing in dwelling provision. Thus the renewal and major repair of dwellings becomes a finance issue which is designed into the provision of new dwellings rather than remaining a liability for future generations. This does not imply that the current balance between building costs, subsidies, rents and design standards has been optimised.

The brief review of housing regeneration in Britain presented above highlights the broad shifts in approach and policy. It also serves to identify important elements of the framework within which housing regeneration occurs.

Strategic Planning Framework

The strategic context for housing investment has not been well developed in the past. Initiatives have not generally been set in the context of multi-functional, multilevel plans. McGregor and MacLennan (1992, p. 11), in a study of Scottish housing regeneration initiatives, found no instance where 'neighbourhood plans or initiatives were linked into a wider area initiative which then meshed into District Housing Plans and Regional Structure Plans in any detailed way'.

Similarly, the HIP process and development of local housing strategies only became effective in England and Wales in the 1980s, driven by the DoE and the Welsh Office. Initially they focused purely on housing issues (need, demand, supply, finance, etc.). It was only as the 1980s progressed that the wider relationships with other strategies such as economic development, planning, anti-poverty and community care have begun to be identified and interlinked with housing (Audit Commission, 1992).

Labour Market Framework

While the general history of urban regeneration and renewal is that it has been housing led, there is widespread acceptance that area renewal cannot simply be housing focused. The 1988 'Action for Cities' regeneration initiative in England was matched in Scotland by the 'New Life for Urban Scotland' White Paper, which identified four partnership areas in which a locally co-ordinated approach to housing-led urban regeneration would be applied (Scottish Office Central Research Unit, 1996). The emphasis in Scotland after 1988 was on a more integrated approach to the economic, social and physical regeneration of (municipal) housing areas (McCarthy, 1995). The continuing need for more co-ordinated, integrated and participative approaches to housing regeneration in England was underlined by the introduction of the Single Regeneration Budget in 1994.

Despite this recognition, housing regeneration initiatives have been singularly unsuccessful to date in linking housing investment and local economic development (McConnachie, Fitzpatrick and McGregor, 1995). Initiatives have been more successful in achieving housing and environmental change than economic and social change. Thus there is the danger that, in areas receiving housing and environmental investment, problems will simply re-create themselves because underlying economic and social conditions remain unimproved.

Community Regeneration Framework

Housing regeneration can not be undertaken without changing the communities in the areas where the investment occurs. However, the social objectives of housing regeneration have not always been clear or even explicitly stated. In this context housing regeneration may be an end in itself or a means to a greater goal of community regeneration.

It is evident that there has been a trend towards greater public involvement in housing regeneration initiatives (McGregor and MacLennan, 1992). This has ranged from consultation and community involvement in the design and planning of housing regeneration programmes, to a formal role in the partnership initiatives. These approaches may, however, not always be consistent with the strategic objectives involved.

In some instances community regeneration objectives are more to the fore, where the emphasis is on retaining, empowering and improving the economic competitiveness of the existing communities. The creation of housing co-operatives in the Scottish Partnership Areas exemplifies this approach (Scottish Office Central Research Unit, 1996).

In other circumstances, the objective may be to achieve sustainable housing regeneration by creating balanced communities. In this situation tenure diversification is as important as community participation and co-ordinating employment initiatives.

Just as coherent strategies for housing improvement have been found to be necessary, so more conscious area or estate strategies are needed in public sector housing management. New approaches to housing management have formed a crucial element in the modernisation and refurbishment of large municipal estates. The most prominent experiments, in England, have formed the Priority Estates Project. The approach adopted emphasised estate-based, local management, permanent local offices, local lettings and a local repairs team, resident caretakers, beat policing, flexible small-scale management, and estate budget, tenant participation, training and continual upgrading of the estate and environment (Glennerster and Turner, 1993; Power, 1987)

Organisational Framework

Housing regeneration has taken place within the context of a range of very different organisational structures and processes. These are reflected in the three approaches described above – the use of administrative mechanisms, the area-based focus and the partnership approach.

It has been a policy of successive governments to implement housing renewal by changing the administrative and organisational structures of housing agencies. This, of course, includes the extension of the role, powers and resources of the Housing Corporation following the 1974 Housing Act. It also includes the creation of new agencies such as Scottish Homes and Tai Cymru following the 1988 legislation. The changing role of housing associations as the main providers of social rented housing, and the consequent change in the role of local housing authorities from landlords to enablers, is now well known. The emergence of housing companies reflects a continuation with this theme of policy which affects both the planning and implementation of housing regeneration. Local housing companies have been developed as a specific model to seek to tackle serious and extensive disrepair in the local authority stock by using private finance, whilst retaining a minority local authority interest.

Housing regeneration initiatives have taken a variety of forms and thus the organisational structures and the organisational issues involved have also varied. It would be possible to identify a typology of housing regeneration initiatives based on factors such as:

- developer: housing association led, local authority led, partnership;
- agency: single agency, multi-agency;
- ownership: single tenure, multi-tenure, tenure change;
- location: inner city, peripheral estate, small town;
- size: defined units (e.g. HAA), single estate, neighbourhood;
- regeneration objectives: housing, multi-sectoral;
- housing objectives: rehabilitation (e.g. HAA), modernisation (e.g. peripheral estate), mixed redevelopment, tenure diversification.

The administrative and organisational issues vary depending on the type of initiative involved. For initiatives where single agencies are involved, or where a range of agencies work separately in the same area, the organisational issues include: area designation (e.g. the definition and boundaries of the Housing Action Area), staff structure, type and scale of staffing. In partnership initiatives the organisational issues will also include issues of co-ordination, targets, programme management, monitoring and contractual liabilities.

The Implementation of Strategy

Chapter 2 identified that successful urban regeneration requires a strategically designed approach with a longer term purpose in mind. It emphasised that it is locally based, 'to recognise and accept the uniqueness of place' (Robson, 1988, p. 102). It should reflect the wider circumstances of the city or region in which it is located and will, by implication, involve a multi-sector and multi-agency approach (Hausner, 1993). It should seek to reduce social exclusion by addressing the needs of social and economically disadvantaged areas (McGregor and McConnachie, 1995).

These broad principles indicate the key elements involved in area-based housing regeneration:

- balanced and self-sustaining communities;
- integration to the wider context of the urban economy and the labour market;
- partnership between agencies, local government, private sector and communities;
- community involvement;
- private investment.

Box 8.1   Upper Dens Project, 1984

The Upper Dens was essentially a housing development, in a mixed commercial and residential area, on a site which had been designated until 1980 for light industrial use. The site contained redundant jute mills and a (largely subterranean) burn. Private developers who were approached were unconvinced of the market for housing in converted mill buildings. The site would require expensive environmental works and there were no grants for private sector housing development. Eventual partners in the project were three housing associations, Dundee City Council, the Housing Corporation, the Scottish Development Agency and the Historic Buildings Council.

The site provided 53 new build family houses, 72 flats in the mill conversion; 36 sheltered housing units; and 53 new build flats for vulnerable single people and 27 supported accommodation units for mentally ill and handicapped people.

The main objectives of the programme were to meet housing need and environmental improvement. There was no community involvement and no community on or near the site. There was no private sector involvement. All the housing provided was for rent.

Subsequently there has been new housing, shopping and commercial development in the adjoining area and seven redundant jute mills have been converted to housing for both sale and rent by public and private sector agencies. The impact of bringing these sites back into active use has been significant in Dundee's urban regeneration and in bringing people back to the centre to live.

The case studies presented in this section (Boxes 8.1, 8.2 and 8.3) are intended to be illustrative of attempts to implement these approaches to housing regeneration. By focusing on one medium-sized city, Dundee, they also illustrate the development of strategy during the last fifteen years.

Box 8.2   Mid-Craigie, 1992

This large 1930 Housing Act slum clearance estate was, almost from its inception, a 'problem estate'. Though substantially sound houses with relatively good space standards, the estate had been subject to several unsuccessful attempts at environmental and housing improvement and modernisation during the 1970s and 1980s, which had been undertaken following extensive tenant consultation. In 1989 the council established an Area Renewal Strategy in response to problems of high turnover and steady population decline. The strategy aimed to: promote stability; improve housing which had a long-term future; and encourage other forms of housing on sites cleared by demolition of surplus properties. The strategy adopted a partnership approach with local community groups, voluntary groups, Scottish Homes and private sector agencies.

Since 1990 almost 800 properties have been demolished. Since 1993 in a joint venture with Wimpey Homes 255 new houses for sale have been started. Tenants have established a housing co-operative which aims to build 180 houses for rent. Capital receipts to the council are being reinvested in improvements to those houses which have a long-term future (102 properties have benefited from new windows and central heating).

Mid-Craigie has been agreed as one of the key areas for the Dundee Partnership's Community Regeneration Strategy and for which Priority Partnership Area status (equivalent to the SRB in England) is being sought.

Mid-Craigie Community Business provided environmental improvements and it is hoped that Scottish Enterprise Tayside will contribute significant funding for recreation areas and environmental improvements.

Box 8.3   Camperdown Works, 1994

This 32-acre site is located to the north-west of the city in an area which consisted of over 80 per cent local authority housing. The site, which is adjacent to a second-tier declining shopping centre (Lochee), formerly contained the world's largest jute works on which production ceased in 1981. It was the largest single site available for redevelopment in the city for many years and was bought by a local developer in 1990. The project had two main components: a complex of leisure and retail facilities with associated parking and an integrated housing development around the axis of the main mill building.

The Camperdown Partnership was established involving the private owner of the site, the Council, Scottish Homes and two local housing associations.

The redeveloped site, completed in 1994, includes a leisure park, retailing, parkland and housing, and has involved new building, conservation of listed buildings, mill conversions and environmental improvements. The leisure park includes a multi-screen cinema, megabowl, nightclub and bingo hall. A Grade B listed building now houses a 38,000 sq. ft superstore. Parkland and children's play areas have been provided around the Cox's Stack built in 1861 in the Campanile style. The housing is in six main phases, four developed by housing associations. Site 1 provides 53 sheltered housing units for rent; Site 2 includes 64 low-rise family dwellings for rent; Site 3 provides 36 three and four person flats for sale in the converted former railway sidings; Site 4A provides 51 two-storey dwellings for shared ownership; Site 4B provides 42 low-cost home ownership dwellings built by Woolwich Homes; Site 5 is the High Mill converted to 74 flatted dwellings for sale, carried out by the developer.   *continued over*

The total cost of the development was around £51 million of which £39 million was provided by the private sector. All the housing projects were grant aided by Scottish Homes at a total cost of £11.2 million for 350 houses and flats. The project has involved new employment in an area of high unemployment, together with new recreational facilities. The housing has led to greater tenure balance in a formerly mono-tenurial area. Additional private sector housing development is taking place in an adjacent area.

Conclusion

As we have seen, the quality of housing, and its surrounding environment, has major social and cost implications. It is also self-evident that housing is, or should be, a long-lasting and durable commodity. An adequate supply of housing, to acceptable modern standards and at appropriate access costs is perhaps the most cost-effective form of infrastructure that can be provided. Housing standards have demonstrable implications for health standards, levels of criminal activity and degrees of educational attainment. If the supply or quality of housing is inadequate there are inevitably heavy cost implications for the providers of social services, often in the form of irrationally expensive emergency solutions such as bed and breakfast accommodation. The private sector has proved very successful in working in partnership with local authorities and housing associations to bring forward very large housing and regeneration projects in very difficult areas. A greater degree of certainty in the application of planning policy coupled with government measures to promote long-term investment in housing will help to provide the stable conditions necessary for the housing industry to respond efficiently to the scale of the projected needs.

Key Issues and Actions

- Housing can be a driver of urban regeneration and it is an essential element of most schemes.
- Clear knowledge of the housing market is an essential pre-condition for good policy.
- Housing needs are difficult to calculate with any accuracy, but a working estimate is essential in order to allow for sites to be identified.
- Good housing helps to improve health, reduce crime and enhance quality of life.
- Social integration can be assisted through housing provision.
- Joint finance and partnership are fundamental to the provision of much urban housing.

References

Audit Commission (1992) *Developing Local Authority Housing Strategies*, HMSO, London.

Ball, M. (1996) *Investing in New Housing*, The Policy Press, Bristol.

Best, R. (1997) City delights, *Town Planning Review*, Vol. 62, pp. 46–7.

Champion, A., Atkins, D., Coombes, M. and Fotheringham, S. (1998) *Urban Exodus – a Report for CPRE*, Department of Geography, University of Newcastle upon Tyne.

Crook, A.D.H., Disson, J.S. and Dark R.A. (1996) *A New Lease of Life: Housing Association Investment on Local Authority Estates*, The Policy Press, Bristol.

Fordham, G. (1995) *Made to Last*, Joseph Rowntree Foundation, York.

Glennerster, H. and Turner, T. (1993) *Estate Based Housing Management: An Evaluation*, Department of the Environment, London.

Gummer, J. (1997) New deal needed, *Town Planning Review*, Vol. 62, Winter, pp. 24–5.

Gwilliam, M. (1997) *Housing and Regeneration – How a Green Shield Levy Can Help*, The Civic Trust, London.

Hall, P. (1998) The Great Housing Land Debate, *Town Planning Review*, Vol. 62, Winter, pp. 42–3.

Hausner, V.A. (1993) The future of urban development, *Royal Society of Arts Journal*, Vol. 141, no. 5441, pp. 523–33.

Jones, H.R. (1990) *Population Geography*, Paul Chapman, London.

King, D. (1993) Demography and house-building needs, in T. Champion (ed.) *Population Matters*, Paul Chapman, London.

Lee, P. and Murie, A. (1997) *Poverty, Housing Tenure and Social Exclusion*, The Policy Press, Bristol.

Lever, W.F. (1994) *The Future Structure of Scottish Cities*, Scottish Homes, Edinburgh.

Llewelyn Davies (1997) *Sustainable Residential Living*, LPAC, London.

London Planning Advisory Committee (1998) *New Face of City Living*, LPAC, London.

McCarthy, J. (1995) Sustainable housing regeneration, *Housing Review*, Vol. 44, no. 6, pp. 122–4.

McConnachie, M., Fitzpatrick, I. and McGregor, A. (1995) *Building Futures*, Glasgow University, Glasgow.

McCrone, G. and Stephens, M. (1995) *Housing Policy in Britain and Europe*, UCL Press, London.

McGregor, A. and MacLennan, D. (1992) *A Review and Critical Evaluation of Strategic Approaches to Urban Regeneration*, Scottish Homes, Edinburgh.

McGregor, A. and McConnachie, M. (1995) Social exclusion, urban regeneration and economic reintegration, *Urban Studies*, Vol. 32, no. 10, pp. 1587–600.

Power, A. (1987) *The Priority Estates Project Experience*, Department of the Environment, London.

Reynolds, F. (1998) *Urban Exodus*, CPRE, London.

Robson, B.T. (1988) *Those Inner Cities*, Clarendon Press, Oxford.

Robson, B. T. (1994) *Assessing the Impact of Urban Policy*, HMSO, London.

Scottish Office Central Research Unit (1996) *Partnership in the Regeneration of Urban Scotland*, HMSO, Edinburgh.

Taylor, J. (1997) City strategy, *Town Planning Review*, Vol. 62, Winter, pp. 17–18.

Thomas, A. (1986) *Housing and Urban Renewal Residential Decay and Revitalization in the Private Sector*, Allen & Unwin, London.

Vann, P. (1996) New Household Projections Taken with a Pinch of Salt, *Planning*, Vol. 1146, 24 November, pp. 24–5.

Watt, K. and Summers, Y. (1995) *Housing in Scotland*, Scottish Homes, Edinburgh.

Wilcox, S. (ed.) (1994) *Housing Finance Review 1994/95*, Joseph Rowntree Foundation, York.

PART 3

KEY ISSUES IN MANAGING URBAN REGENERATION

9 Regeneration by Land Development: the Legal Issues*

Amanda Beresford, Richard Fleetwood and Mark Gaffney

Introduction

The legal aspects of a property regeneration project are diverse, involving numerous statutory (embodied in written laws passed by Parliament) and common (unwritten laws evolved from principles established through decisions issued by the courts) law provisions. Specialist legal advice will usually be needed on the topics of commercial property, environment and planning. It may also be required in other areas such as tax and construction. The early identification of the relevant requirements and implications of the law will usually enable a project to proceed in the most efficient manner.

This chapter can do no more than highlight some of the main legal issues which a particular property regeneration project may have to address. What follows, therefore, is not a comprehensive consideration of each relevant area of law, but an introduction to some of the main areas that will need to be considered. Additional reading is suggested in the Further Reading section at the end of the chapter.

This chapter examines the key issues and the main relevant provisions of the following specialist legal practice areas:

- legal structures for delivery;
- commercial property law;
- environmental law;
- planning law.

Legal Structures for Urban Regeneration

The vehicle or organisation for regeneration will need to be considered. Often regeneration, particularly in the case of a smaller project, may be carried out by an existing private development company. However, it may

*The Legal Issues in this chapter are as relevant to England and Wales only.

be necessary to set up a new vehicle or to utilise one or more of the existing organisations charged with regeneration responsibilities.

This section examines the different legal structures and approaches that can be used to pull together an urban regeneration scheme and considers some of the public sector parties who may need to be involved. It is broken down into the following subsections:

- Legal structures:
  - a limited company (limited by shares or by guarantee);
  - other joint ventures or partnership arrangements where the parties agree, contractually – for example, through a development agreement – to undertake a particular scheme;
  - a charitable trust.
- Existing or proposed organisations:
  - English Partnerships;
  - Regional Development Agencies;
  - local authorities.
- The Private Finance Initiative.

The appropriate vehicle will depend, to a large extent, on the views and competences of the key parties who will need to be involved in the project and may include a consideration of the role of landowners or planning authorities. The majority of urban regeneration schemes involve a number of key parties and this chapter assumes that there will be at least two parties involved (for example, a private sector and a public sector party). If the project only involves one company or entity then, in all likelihood, that party will own the site being regenerated and will contract with third parties for any work required on the scheme, including construction or site clearance works.

Legal Structures

Using a Limited Company

Limited companies are of two main types – limited by shares or by guarantee. Companies limited by shares are by far the most common. Shareholders take shares in return for providing assets or paying money to the company for those shares. There is no limit to the number of shareholders who can be involved. Shareholders can agree between themselves through the company's articles of association and/or through a shareholders' agreement how the company is to be run (for example, by majority decisions at the board or shareholder level except for a reserved list of 'veto' key matters) and how any profits are to be shared, which need not follow the voting rights of the shareholders.

Limited liability companies give the shareholders the comfort of limited liability in relation to third parties dealing with the company. This means that if the company were to become insolvent then, except in limited circumstances, neither the shareholders nor the directors of the company would be liable to third parties for the debts of the company.

On the other hand, guarantee companies do not have shares or share-holders – they have members. The rights of persons to become members are set out in the company's articles of association. Guarantee companies are often favoured in relation to grant financed or assisted projects as a guarantee company cannot distribute profits to its members. Even on the winding up of a guarantee company, any surplus assets (after the payment of creditors, etc.) would, in the normal course, be reapplied to another guarantee company, charity or trust with the same or similar objects to the original guarantee company.

In the remainder of this chapter companies set up specially for a particular project (whether limited by shares or guarantee) are, adopting the jargon, referred to as special purpose vehicles or SPVs.

Joint Venture or Partnership Arrangements

The cost (both in terms of the initial set-up and the ongoing running costs) of setting up a company or using an existing company for a project will not always be justified. Furthermore, a company is not always needed as a vehicle through which to undertake a regeneration scheme. The main alternative is a joint venture or contractual arrangement whereby the various parties to the project agree – without setting up a new company – how they are to undertake the project, what contributions (in money or in kind) each party is to make and what the financial entitlement of each of the parties is to be.

If the scheme is to be taken forward on a partnership approach, then the purpose of the partnership will need to be clearly identified as will the roles and responsibilities of those involved. Particular care needs to be taken on financial matters in terms of obligations to fund the scheme and the rights of the participants to share in any profits or surplus at the conclusion of the scheme.

Box 9.1 summarises some of the main differences between corporate and non-corporate structures. As a general rule, a corporate SPV should only be used where the complexity of the project justifies the time and cost that is entailed.

Charitable Trusts

There may be advantages in setting up a charitable trust to undertake a particular scheme, although the more limited commercial freedom enjoyed by a charitable trust needs to be weighed against the benefits (principally tax related) of charitable status. It is important to note that, in order to obtain charitable status, it is necessary to demonstrate that the trust is established for charitable purposes, which include the relief of poverty, purposes beneficial to the community, the advancement of education, and public recreational purposes.

Box 9.1   Key differences between corporate and non-corporate structures

Corporate SPV	*Non-corporate*
Members of the company are insulated from liability to third parties (but consider the public relations implications of being involved with a company which becomes insolvent). It is important to note, however, that unless the SPV has a strong balance sheet, third parties are unlikely to deal with it without direct guarantees from members/shareholders	Those involved will have to contract direct with third parties in their own names and will be exposed to liabilities accordingly
Can own assets/land in its own name (this can be a particular advantage in ring-fencing assets from members/shareholders and third parties and minimising interference from members/shareholders); this also assists in land/site assembly	Parties will have to keep ownership of assets. This may give rise to operational difficulties or interference from the parties
Will be subject to corporation tax regime (this includes flexibility for surrender of losses by the SPV to shareholders or vice versa which may be advantageous)	Each party will be subject to tax in its own right. Arrangement could constitute a partnership in which case it will be necessary to consider: – tax regime for partnerships – joint and several liability of partners
Provides a familiar legal entity into which third parties can invest	Third parties will need to contract directly with all of the parties involved
Provides a forum for decision-making and management (through the SPV's board of directors)	No forum – parties need to agree between themselves how the venture is to operate
SPV will be a separate legal entity with statutory obligations to keep accounts, etc (this is an advantage in terms of providing certainty; but a disadvantage in terms of cost)	No entity to hold assets or profits generated. These will have to be dealt with contractually
Enables the 'branding' of a specific development proposition to the outside world	
	Partnership approach may be favoured in the context of competing for some sources of grant funding (e.g. Single Regeneration Budget funding). Allows for a wider 'community' involvement

Existing Organisations Involved with Urban Regeneration

English Partnerships
English Partnerships or, to give it its statutory title, the Urban Regeneration Agency, was set up under the provisions of Part III of the Leasehold
Reform, Housing and Urban Development Act 1993 (the 1993 Act). It
came into full operational effect on 1 April 1994, taking over the functions
of English Estates, City Grant and Derelict Land Grant. The 1993 Act sets
out the objects of EP which, in outline, include the securing of regeneration
of land in England which is unused or ineffectively used, or which is contaminated, derelict or unsightly. In order to achieve these objects, EP has
wide-ranging powers (see Section 160 of the 1993 Act). The head office is
in London, and the organisation has a network of regional offices around
the country, including two corporate offices at Haydock and Gateshead.

In the main EP pursues its objects through grant funding projects which
are brought to it by both the public and private sectors. The most common
means of funding is through the provision of gap funding for development
agreements, whereby EP will bridge the 'gap' between the costs of a project – for example, constructing a building on reclaimed land – and what
the finished building is worth on the open market assuming the finished
building is worth less than the cost of the works (hence the 'gap'). A
standard approval process exists, which must be followed by applicants, on
the basis of which EP will decide (having regard to the outputs generated,
including land reclaimed, or jobs created or secured) whether grant funding should be provided and, if so, on what terms. In addition to gap funding, EP is also able to provide loans and guarantees. In some cases EP will
participate in joint ventures. English Partnerships has been an important
agent for change during recent years and it is expected that much of EP's
accumulated experience and expertise will be transferred to the new Regional Development Agencies.

Regional Development Agencies
Regional Development Agencies are a creation of the current Labour
government with their main promoter being the Deputy Prime Minister,
John Prescott. The nature, role, and proposed powers of RDAs were the
subject of consultation soon after the general election of May 1997. The
RDAs will incorporate the roles of English Partnerships and the Rural
Development Commission in the regions and will also co-ordinate the
activities of a range of other organisations. Nine 'shadow' RDAs came into
operation in the autumn of 1998, with the fully-fledged RDAs being operational from April 1999. Further details of how RDAs are to operate, how
they will be staffed and what their drivers are likely to be, will become
apparent during the remainder of 1998. However, the key objective of
RDAs will be to tackle regeneration by providing effective and properly
co-ordinated regional development. It is also anticipated that RDAs will

co-ordinate land assembly, physical and economic regeneration, support for small businesses and the promotion of inward investment.

Local Authorities

Local authorities usually have a key role to play in urban regeneration schemes, sometimes as a landowner, sometimes as the relevant planning authority and sometimes as both.

A key point to consider at an early stage in any regeneration scheme involving a local authority is how the transaction or scheme will be treated by the local authority in terms of the local authority's accounting requirements. In outline, legislation was introduced during the 1980s aimed at curtailing overspending by local authorities at the expense of local tax-payers (through rates, the community charge or council tax) or the national purse (through payments financed by tax or government borrowings). The relevant legislation introduced an extremely complex web of requirements and restrictions relating to both capital and revenue expenditure and the receipts of local authorities. In the late 1990s there have been some relaxations to the various rules (some are linked with the Private Finance Initiative for which details are provided below).

At the heart of the local government 'capital finance regime' is Part V of the Local Government and Housing Act 1989 (the 1989 Act). This specifies when a local authority will be deemed to control a company so that, in effect, the finances of the company (whether assets or liabilities, but particularly liabilities) must be treated as part of the overall financial position of the local authority.

In addition to the 1989 Act, regard must also be had to the Local Government (Companies) Order 1995 (the 1995 Order), which brings a range of interests of local authorities in companies at below control level into the regulated company net for capital finance purposes. It is the 1995 Order which gives rise to the situation in which local authorities often restrict their shareholdings in companies to less than 20 per cent. The key point to be considered at the outset in the structuring of any transactions in which a local authority is involved, is that, through careful planning, a local authority may be able to secure favourable treatment for its involvement in the project for capital finance purposes. For example, if a local authority transfers land to a non-regulated SPV in return for shares or loan stock in the SPV, then no capital receipt will arise for the local authority. On the other hand, if the land were to be transferred by the local authority in return for cash, the local authority would be obliged, under the capital finance regime, to 'set aside' (in normal circumstances) 50 per cent of the receipts to reduce its borrowings (75 per cent of the proceeds in the case of housing asset sales).

There have been a number of recent changes to the legal regime relating to local authorities. Key points to note are:

• the proposed release by the government of capital receipts of up to £5 billion over a five-year period for use towards housing and housing-related regeneration (see the Local Government Finance (Supplementary Credit) Act 1997);

- incentives to local authorities to undertake 'Design, Build, Finance and Operate' type contracts (see the Local Government (Capital Finance) Regulations 1997);
- a new self-certification scheme for local government PFI schemes which guarantees private sector partners (including banks) that they will be compensated if a contract is ruled unlawful (see the Local Government (Contracts) Act 1997);
- from September 1998, extensive relaxations to the capital finance regime so that, in many cases, local authorities will be able to dispose of certain surplus assets without any requirement to 'set aside' part of the monies received.

In addition to taking account of the capital finance rules relating to a local authority, a private sector party should take steps to satisfy itself that:

- the local authority has an express statutory power to become involved in the scheme in the manner envisaged (the key general power relied upon by local authorities is contained in Sections 33 to 35 of the 1989 Act relating to the promotion of economic development of a local authority's area in accordance with an approved annual plan);
- the express power has been properly exercised (in terms of the local authority considering the scheme at appropriate officer and member level);
- the relevant legal agreements have been properly executed in accordance with any minuted authorities and/or standing orders.

Obviously, the Local Government (Contracts) Act 1997 will provide significant comfort to private sector partners.

The Private Finance Initiative

The Private Finance Initiative was launched in 1992 with the aims of improving the quality and quantity of public sector capital projects, and of delivering high-quality and more cost-effective public services. It attempts to do this through encouraging partnerships and by involving the private sector more directly in asset provision and operation.

The present government has confirmed its support for PFI and believes it should be the route of choice for all public sector procurement where it can deliver superior value for money. A wide range of PFI schemes have been agreed to date with many more under negotiation. However, PFI requires, as a key ingredient, the existence of an asset (for example, a school, a road, a hospital or a prison) which the private sector can design, build, fund and operate. Consequently, PFI will not have a role in urban regeneration unless the regeneration scheme involves the construction of new facilities of a type capable of being funded through PFI. This is an area where there

is significant overlap between the overall PFI regime and the capital finance regime applicable to local authorities. A detailed review of the way
PFI operates (particularly in the context of local authorities) is outside the
scope of the basic introduction provided by this part of the chapter. Should
you require further information in this area you should read the Department of the Environment, Transport and the Regions publication *Local
Government and The Private Finance Initiative – An Eplanatory Note on
PFI and Public/Private Partnerships in Local Government* was published in
February 1997 and updated in September 1998.

Property Law

The property aspects of any regeneration scheme can be generally divided
into two parts: first, the site and how to assemble the same and, second, the
influence that any third party rights over the site will have on the proposed
development. These issues are made more complex in the case of urban
regeneration, as opposed to the development of a greenfield site, in that a
site in an urban area is more likely to be in the ownership of a number of
parties and to be subject to a greater number of third party rights.

This section of the chapter examines two major issues: first, the assembly
of a site and how a developer may gain control of a site and, second, the
impact of third party rights on the proposed development and what can be
done to remove such encumbrances.

Site Assembly

The developer, which may be a SPV as previously mentioned, will first
have to establish how many and what freehold and leasehold interests need
to be acquired to assemble a site, and who owns those interests. A search at
the Land Registry will identify the freehold and leasehold interests which
are registered and will also give the name and address of the owners of such
interests. However, the search will not reveal interests which have not yet
been registered (for example, because there have been no dealings in
relation to the land since compulsory registration was introduced) nor will
it give details of leasehold interests for a term of less than 25 years as these
are not registrable. The proposed developer will have to rely on making
enquiries of any occupiers of the land to try and establish the ownership of
the freehold and any leasehold interests.

There may well be parts of a proposed site for which the developer is
unable to establish who has the paper title to the land and no one can
establish title by way of adverse possession (by exclusive possession for 12
years or more). A common example of this in urban regeneration is

provided by access roads at the rear of properties which have never been adopted. To protect against a landowner subsequently claiming the title to the land after the development has commenced and bringing an action in trespass for damages and/or an injunction to prevent the development proceeding, the developer has the following options:

- to seek to design the scheme so that the piece of land in question does not form an important part of the scheme;
- to obtain effective title indemnity insurance against such a claim;
- to seek the co-operation of the local authority to use its compulsory purchase powers to acquire the land in question.

Having identified the interests in the site which the developer needs to acquire, consideration should be given to how the developer can secure those interests. There are a number of different forms of agreement which can, to a greater or lesser extent, give the developer control of the site as follows:

- an unconditional contract;
- a conditional contract;
- an option agreement;
- a pre-emption agreement.

A developer is unlikely to want to proceed on the basis of an unconditional contract to purchase the land, unless the land to be sold comprises the entire site and either has outline planning permission for the proposed development or the developer is confident that he can obtain the necessary planning permission.

A conditional contract will allow the developer the comfort of having the ability not to purchase the property if certain conditions are not satisfied such as planning and pre-lets being secured and the environmental condition of the land being satisfactory. The greater the number of uncertainties in respect of the development proposals, the greater the flexibility a developer requires and is, therefore, more likely to seek an option which, in effect, gives the developer total discretion as to whether or not to proceed with the acquisition.

Of course, whether the developer can negotiate a conditional contract or option agreement will depend upon the attitude of the seller, and whether the seller believes its interest would be bettered by selling the site immediately. The landowner is probably more likely to agree to enter into a conditional contract or option agreement if the period within which any condition has to be satisfied, or option exercised, is reasonable and the seller receives some financial recompense for entering into the agreement which, if the sale proceeds, will form part of the purchase price but if it does not will be retained by the land owner.

With a conditional contract or an option agreement the seller may, in return for the uncertainty on the sale of his site, either consider a higher price or consider some involvement in the development. The simplest form of involvement is by way of a profit share. The developer needs to be careful as to how the rights to future profits are secured as they may in turn impact upon the developer's ability to finance the development.

The final form of agreement referred to is a pre-emption agreement. This merely gives the developer a right of first refusal should the seller decide to dispose of his interest. This is unlikely to be attractive to a developer, especially in relation to an important element of the site. However, the developer may use a pre-emption agreement in relation to any area of land which, for example, could be used in the future for the expansion of a proposed scheme.

In respect of all the above agreements, in order to protect the developer's rights under the same they need to be registered on the seller's title. Failure to register the agreements means that the land could be sold to a third party and, whilst the developer would have a claim in damages against the landowner, the agreement could not be enforced against the third party purchaser.

If the developer is having difficulties in agreeing a deal with a landowner then, especially in the cases of urban regeneration, the developer may seek the assistance of a local authority to use, or at least threaten to use, its compulsory purchase powers to assist in the negotiations with the uncooperative landowners. It should also be remembered that English Partnerships have the power upon authorisation from the Secretary of State to compulsorily acquire land under the provisions of Section 162 of the Leasehold Reform Housing and Urban Developments Act 1993. If compulsory purchase powers are used to assemble a site, this can cause delays in the developer's programme. The developer may also need to recompense the local authority or English Partnerships for the costs incurred in exercising these powers, which may cause the developer problems with cash flow.

In any documentation providing for the local authority to assist the developer in assembling a site, the relevant provisions to be enforceable must be worded so as not to fetter any statutory rights or obligations of the local authority.

Third Party Rights

The development of a site has the potential to interfere with third party rights. If the proposed development interferes with the rights enjoyed by third parties, this can give rise to an injunction to stop the development or may result in a claim in damages. It is important that the title to a site is investigated as early as possible. In the case of a title which is registered, this can, in part, be carried out without the need for any co-operation from the landowner as details of the title are available at the various District Land Registries. However, in the case of a title which is unregistered, the developer will be unable to examine the title deeds and documents without the co-operation of the landowner.

Easements

Easements are rights benefiting one property over another. Easements may be created by statute, expressly, impliedly or by presumed grant or prescription (for example, by reason of the fact of previous long use of a right). Therefore, merely examining the title documents will not necessarily reveal all third party rights. An inspection of the site will also need to be undertaken to see if any such rights are apparent and enquiries will have to be made of the seller to establish whether any such rights exist.

A site subject to easements (for example, rights of way or for service media) may in effect be sterilised in respect of those parts of the site so affected, and will make the development extremely difficult unless the developer is able to plan the development 'round' them. If an easement is identified, the developer's solicitors first need to consider if the easement is, in fact, enforceable. If the easement is enforceable, it may be necessary to:

- negotiate a release with the owner of the right;
- obtain effective title indemnity insurance;
- seek the co-operation of the local authority to exercise its compulsory purchase powers in respect of the easement;
- appropriation – that is, if part of the site is affected by a covenant, is, or has been, in the ownership of the local authority and has been appropriated for planning purposes, then the right may be converted into a claim for compensation rather than entitling the person with the benefit to seek an injunction; indemnity insurance will be required to cover the cost of the compensation, but the developer has the comfort that an injunction cannot be obtained.

Restrictive Covenants

These are restrictions on a title which limit what can be done on a site. It may be necessary to consider the rules governing the enforceability of covenants to see whether a covenant is enforceable and if so by whom. It is not always clear who has the benefit of a restrictive covenant. Unlike the burden of a covenant, the right is not usually registered on the title of the land having the benefit of it, and, especially with older covenants, the developer may take a view that a covenant will not be enforced. The developer should bear in mind that it will be necessary to convince prospective tenants, fund providers and purchasers that a covenant is unenforceable.

Again, if a covenant is enforceable and the developer is unable to design the scheme 'round' the restriction, there are a number of options open to a developer as follows:

- seek to negotiate a release with the party having the benefit of the covenant;
- obtain defective title indemnity insurance;
- apply to the Lands Tribunal for a release or modification of the covenant;

- appropriation as described above;
- seek the co-operation of the local authority in order to use its compulsory purchase powers to acquire the benefit of the covenant.

Public Rights

It is not uncommon in urban regeneration projects for there to be a need to either close or divert a public highway. There are two procedures for closing or diverting a public highway:

- an application to the Magistrates Court under the Highways Act 1980; this can only be done with the co-operation of the local authority and the magistrates may be reluctant to grant an order where an application can be made under the Town and Country Planning Act (see below);
- where there is an existing planning permission in the case of development which would require the closure or diversion of a public highway, then an application can be made to the Secretary of State for the necessary closure or diversion under the Town and Country Planning Act; the problem with this procedure is that in the event an objection is received, an inquiry has to be held which can delay a scheme.

It should be remembered that the closure or diversion of a public highway will not extinguish any private rights which existed. These will have to be dealt with separately.

Environmental Law

A property regeneration project will often give rise to the need to deal with a number of issues governed by environmental law. The main issues relate to:

- waste management;
- contaminated land.

Waste Management

A property development urban regeneration project frequently produces waste in the form of old construction material, excess earth, etc. If the area to be redeveloped has previously been in industrial use, the waste may contain a degree of contamination. The management of such waste, even the temporary disposal of such waste within the site or movement from one part of the site to another, requires compliance with a number of environmental laws. Of particular relevance are laws relating to:

- the definition of waste;

- waste management licensing requirements;
- the statutory duty of care in relation to waste;
- Landfill Tax.

Care should also always be taken to ensure that the development does not result in waste being kept, treated or disposed of in a manner likely to cause pollution of the environment or harm to human health. Breach of this requirement would result in the commission of an offence under the provisions of the Environmental Protection Act 1990.

Depending upon the circumstances, liability for breach of the laws relating to waste can attach to a landowner, developer, contractor or other parties involved in the project. It is, therefore, usually in the best interest of all parties to ensure that any waste is properly dealt with and that responsibilities relating to it are clearly set out in any contractual arrangements.

The Legal Definition of Waste

It is important to establish whether or not what is being dealt with is legally defined as waste. The definition of waste is a complex area of the law. Particular areas of difficulty arise where waste is sold or given to someone else to reuse, or is temporarily stored on the site for reuse elsewhere later in the development project or is treated on site. Failure to correctly recognise that what is being dealt with is waste may result in the commission of an offence. Further, subsequently complying with the law may jeopardise the project's completion time and budget due to unexpected licence fees and the time taken to obtain the necessary permissions, licences and/or consents to deal with the waste. Consideration of this aspect of any proposed development is essential at an early stage.

Waste Management Licensing

If waste is produced, kept, treated, disposed of, or is subject to some recovery operation, then authorisation will be required in the form of a Waste Management Licence granted by the waste regulatory authority, which in England and Wales is the Environment Agency. Applications can take some time to process, may require a significant amount of supporting information, and a fee is payable.

There are a number of exceptions to this general rule that are a consequence of the Waste Management Licensing Regulations 1994. These regulations contain a lengthy list of activities involving waste which expressly do not require a waste management licence. Of particular relevance in this context may be exemption number 19, which exempts the storage, or use on site, of certain construction wastes for specified construction works, and exemption number 9, which provides that the spreading of certain construction or demolition wastes on land in connection with specified reclamation or improvement is exempt. Both of these exemptions may need to be considered in detail. Some of these exempted activities still require registration with the waste regulatory authority.

Duty of Care

The Environmental Protection Act 1990 imposes on everybody who produces or deals with most wastes a legal duty of care. Basically, this involves a duty to prevent waste:

- causing pollution of the environment or harm to human health;
- escaping;
- being transferred to another person who is not authorised or without a proper written description of the waste.

Breach of this duty of care is an offence. Discharging the duty of care requires attention to such things as storing and packing waste properly, describing clearly what it consists of, dealing only with an authorised carrier, providing the carrier with an accurate transfer note and taking steps to ensure that the waste is ultimately disposed of correctly. Waste arising from a project should, therefore, always be dealt with in accordance with this statutory duty of care.

Landfill Tax

The disposal of wastes to landfill attracts a payment of Landfill Tax. However, there is an exemption (currently under review) relevant to property development regeneration projects dealing with previously contaminated land – the historic contaminated land exemption. If the exemption does not apply, the cost of the development will have to take account of the tax. If the exemption does apply, it is important to claim the exemption from the Inland Revenue at least 30 days before the disposal of the waste takes place. For the exemption to apply there are a number of requirements and these should be considered in detail. HM Customs and Excise publish a relevant information note (1/97 although a revision of this has been indicated). Briefly, the requirements include:

- there must be reclamation of contaminated land which is, or is to be, carried out with the object of facilitating development, conservation, the provision of a public park or other amenity, or the use of the land for agriculture or forestry, or, if none of these is to be carried out, with the object of reducing or removing the potential of pollutants to cause harm;
- reclamation must involve clearing the land of pollutants that are causing harm or have the potential to cause harm;
- the cause of pollution must have ceased;
- the land is not subject to a remediation notice;
- the reclamation constitutes or includes clearing the land of pollutants which would (unless cleared) prevent the land being put to the intended use.

Contaminated Land

Inevitably in many urban regeneration property development projects, it may be that all or part of the site will be contaminated by a previous use. However, it should be borne in mind that dereliction and evidence of previous use are not, in themselves, evidence of contamination. Some derelict and previously used sites (often known as 'brownfield' sites) are not contaminated and many would not fall within the statutory definition of contamination relevant to remediation notices. Indeed the present government is keen to promote the development of 'brownfield' sites and has proposed that a tax may be applied on 'greenfield' sites in future in order to encourage the reuse of 'brownfield' sites. The legislative provisions relating to contaminated land mean that it is essential that any prospective purchaser or developer of land investigates whether or not it is contaminated, as in certain cases, as will be appreciated from the information below, liability can pass with the land. This may involve the appointment of environmental consultants to carry out an environmental investigation and it is important to seek advice as to the appropriate terms and conditions of their appointment. If a site is contaminated, then the principal mechanism for dealing with the contamination as part of a redevelopment project is the planning process. The assessment of the existence of contamination, and the requirement to deal with it, should be dealt with by the local planning authority in the consideration of any application for planning permission and the imposition of any conditions attached to a permission. This is discussed in further detail in the planning section of this chapter. Notwithstanding this, anyone involved in land which may be contaminated needs to be aware of the implications of the various legislative provisions designed to deal with that contamination. The main provisions are:

- statutory nuisance;
- remediation notices;
- works notices;
- civil claims for damages or compensation.

Statutory Nuisance

Certain specified circumstances are held to be a statutory nuisance in re-spect of which a local authority can require rectification through the service of an abatement notice. Failure to comply with an abatement notice is a criminal offence. The specified circumstances include such things as:

- premises in a state prejudicial to health or a nuisance;
- any accumulation or deposit which is prejudicial to health or a nuisance;
- noise;
- dust.

Such circumstances could potentially arise during a regeneration project, either as a direct result of the actual activities during development, or because the activities disclose unknown contamination.

Compliance with an abatement notice can be required of the person responsible for the nuisance or, if that person cannot be found, the owner or occupier of the premises. The implication for an urban regeneration project is that the owner or occupier of the site potentially falls liable under these proceedings to remedy any statutory nuisance caused by a previous owner or occupier (for example, land contamination) who can no longer be found. Also, if the development works themselves cause a statutory nuisance, liability will arise. The intention is that in respect of contaminated land, the relevant statutory nuisance provisions will largely be superseded by the proposed regimes relating to Remediation Notices and Works Notices referred to below.

Remediation Notices

New statutory provisions will require the regulatory authority to identify land which is contaminated and needs attention, and then to serve a notice on the person liable to clean it up; non-compliance would be a criminal offence.

Those involved in property developments will need to assess whether or not a remediation notice is likely to be served in respect of the site in question. The relevant regulations are currently in draft form and the following account is of the situation that would result if they were adopted in their present form. The provisions will be retrospective.

The statutory definition of contaminated land introduced by the Environment Act 1995 is relevant. The result of this appears to be that if there is no harm or significant possibility of harm, land is not contaminated within the meaning of the relevant statute, despite the presence of harmful matter. Land is contaminated only if the harm, or risk of harm, to the non-aquatic environment is significant, or if there is any risk of water pollution. The result is that, probably, the provisions will only affect the most severely contaminated sites.

Where the regulatory authority does identify such contaminated land, a remediation notice can be served requiring remediation works. Failure to comply is a criminal offence. The principal person upon which the notice must be served is the person who caused, or knowingly permitted, the contaminating substance to be in, on, or under the land in question. However, if that person cannot be found, then the notice can be served on the owner or occupier for the time being of the contaminated land. The term 'knowingly permitted' means care has to be taken in any contractual arrangements. It is possible that a funding institution could be held liable as someone responsible for causing or knowingly permitting the presence of the contamination in certain circumstances. Dealing with the concerns of a funding institution in relation to this matter can be an important part of a regeneration project.

There are a number of complex rules that provide exclusions to the general regulations governing liability. These serve to exclude a particular person who would otherwise be liable but only operate when there is more than one person in the particular category, that is, the causing or knowingly permitting category, or the owner/occupier category. The most relevant issues in a property development situation include:

- land sold with information;
- payments for remediation works to another party;
- rack rent leases.

The importance to a regeneration project is that a proper assessment of whether or not the land is contaminated must be made early in the project so that unexpected risks or costs are avoided. Attention to the details of the appointment of environmental consultants will be required. Any remediation works should be to a standard which will satisfy the regulatory authorities, and liability for any residual risk should be addressed in the contractual arrangements between the parties.

Works Notices
If a regeneration project causes, or is likely to cause, pollution of water, as a result perhaps of disturbing contamination on the land, then regulations, yet to come into force, will enable the Environment Agency to serve a works notice on the person who caused or knowingly permitted the pollution. This notice requires the pollution to be cleaned up. Failure to respond to a notice is a criminal offence.

Like remediation notices, works notices mean that knowledge, adequate remediation, attention to the details in appointing environmental consultants, and arrangements to deal with residual liability, may all be important matters to be dealt with in a regeneration project.

Civil Claims for Damages or Compensation
If land is contaminated and the contamination escapes and causes harm to a third party this may give rise to claims from the third party for damages and/or compensation. It is possible for those who have been involved with ownership or development of the land to be implicated in such claims. It is therefore important for this reason also to identify at an early stage whether or not contamination is an issue.

Planning

Generally speaking, planning law requires planning permission to be granted before most forms of development can take place. Regeneration will involve a significant amount of development, which will usually require planning permission and may also require other permissions, for example, listed building consent. It is important at the outset to identify the parts of a proposed regeneration project which require planning or other permissions, and then to obtain all the permissions in an acceptable form. This may involve the completion of planning agreements or pursuing appeals, which will involve a public inquiry. This section, therefore, looks at the following:

- planning permission;
- some other permissions;
- special planning areas.

Planning Permission

Inevitably, any urban regeneration project that involves property development will require planning permission as it will usually involve building, engineering, mining or other operations or/and the making of a material change in the use of any buildings or land. Some forms of (generally minor) development are, in effect, granted planning permission automatically under the provisions of the General Permitted Development Order 1995. These are known as permitted developments. If there is an existing use or development which is not authorised by a grant of planning permission, provided it has existed for the appropriate time (4 or 10 years depending on the type of development) it can be said to have become lawful and a certificate of lawfulness of existing use or development can be obtained from the local planning authority. A similar certificate can be obtained in relation to a proposed use or development if there is doubt as to whether a proposed use complies with what is the legitimate planning use of the site. Applications for either certificate may require detailed evidence as to the past history of the site, and such evidence should usually be compiled in collaboration with a specialist planning lawyer. Overall, the main planning concerns at the beginning of a project will be to identify all of the aspects of the development which require planning permission and to decide whether planning permission is likely to be granted.

Planning law requires the local planning authority to take into account material considerations in determining planning permissions. Material considerations include, among other things, the development plan and planning policy guidance notes. It is particularly important to consider whether a proposed development is consistent with the provisions of the development plan.

Development Plans

Local planning authorities are required to prepare development plans for their area. These may be structure plans, covering strategic matters, local plans, translating the strategic policies into specific policies and proposals and, in Greater London and the metropolitan areas, unitary plans, which are intended to perform the functions of both structure and local plans in these areas. In determining whether or not planning permission is likely to be granted, the development plan provisions are particularly important because Section 54A of the Town and Country Planning Act 1990 requires that the determination of planning applications must be made in accordance with the development plan unless material considerations indicate otherwise. Landowners and developers have an opportunity to influence the provisions contained within development plans through the plan preparation procedure. This procedure requires local planning authorities to undertake a consultation exercise during the preparation of plans and also enables objections to be made to a proposed plan (called the deposit draft) which, if not met by amendments to the proposed plan, can be aired at a public inquiry before an inspector appointed by the Secretary of State. The inspector will then recommend to the local planning authority whether or not any objections should be met through amendments to the plan before it is adopted. Influencing development plans in this way requires a degree of long-term planning on the part of any landowner or developer, since the plans frequently take many years from conception to adoption.

A local planning authority is not bound to comply with the Secretary of State's recommendation in this respect. However, failure to do so runs the risk of the local authority being unable successfully to defend an appeal against a subsequent refusal of an application for planning permission which is consistent with the Secretary of State's recommendations. The importance of any provision in an emerging development plan in relation to the determination of any applications for planning permission, increases the nearer the plan is to adoption.

Planning Policy Guidance

Central government's policies on planning are contained in Circulars and Planning Policy Guidance Notes (PPGs) and local planning authorities will have regard to these in determining any application for planning permission. Box 9.2 identifies some of the most important PPGs.

In mid-1998 the Department of Environment, Transport and the Regions announced that a fundamental review was to be undertaken of planning policy, law and regulation. It is anticipated that the results of this review will be made known during the first half of 1999.

As well as national planning policy guidance, the government also issues Regional Planning Guidance, which it prepares after consultation with local planning authorities and which takes account of national planning policies. Regional Planning Guidance provides the strategic framework within which structure, unitary and other development plans are prepared.

Planning Applications

An outline planning application will usually be the most appropriate first step in the case of a complex scheme involving a change of use as it will

---

Box 9.2   Some Important PPGs

PPG 1 – General Policy and Principles	This sets out the general principles for the operation of the planning system, including the determination of planning applications.
PPG 2	National planning policy in the green belts.
PPG3 – Housing	General policies in relation to housing, affordable housing, housing land availability and new settlements.
PPG4 – Industrial and Commercial Developments and Small Firms	The role of the planning system in relation to industrial and commercial development.
PPG5 – Simplified Planning Zones	The general nature and role of SPZs.
PPG6 – Town Centres and Retail Developments	Including the sequential approach to selecting sites for development for retail employment leisure and other key town centre uses.
PPG12 – Development Plans and Regional Planning Guidance	Government policy in relation to the Development Plan process.
PPG15 – Planning and the Historic Environment	Comprehensive advice on controls for the protection of historic buildings and conservation areas.
PPG16 – Archaeology and Planning	Policy on archaeological remains on land.
PPG23 – Planning and Pollution Control	Including advice on issues relating to contaminated land and waste.

Note: There are other PPGs and circulars which may be relevant to particular projects.

---

establish the principles of that form of development without incurring the cost of working out the final details of the scheme, some or all of which can remain outstanding as reserved matters. An outline permission cannot be implemented until full permission for the reserved matters has been granted.

Usually, the local planning authority has eight weeks within which to determine an application for planning permission. However, it will take longer if complex issues are involved and often the eight weeks is regarded only as a target. If the local planning authority refuses to grant planning permission, grants it with unacceptable conditions or fails to determine it within eight weeks, an appeal can be lodged. The determination of the appeal may involve a public inquiry. Decisions resulting from an appeal can take many months to obtain. It is, therefore, important at the outset to obtain a realistic planning appraisal of the proposed project.

Some types of development have to be referred to the Secretary of State before they can be granted permission by the local planning authority. An example would be a development which is inconsistent with the provisions of the development plan. The Secretary of State can call the application in for his determination and as this may involve a public inquiry, this can significantly delay the determination of the planning application. It is therefore important to identify at an early stage any application which will have to be referred to the Secretary of State.

A planning application need not be in the name of a landowner, but if it is not, then a certificate must be served on the landowner. Only an applicant can lodge an appeal against a planning permission. If a Section 106 Agreement or obligation is required (see below) it will be necessary for the landowner to enter into such an agreement. It is, therefore, important that if the application is made in the name of the developer, arrangements for co-operation between the landowner and any developer in progressing the planning application are addressed in the contractual arrangements.

Section 106 obligations are frequently entered into to facilitate a grant of planning permission. These are agreements or unilateral undertakings by the developer to carry out what is known as planning gain which may, for example, include a restriction on the use or development of the land, the carrying out of specified operations or the payment of a sum of money. Such an obligation is binding on subsequent land owners. A typical example of an obligation in such an agreement is an agreement to pay for the cost of road improvements required in order to accommodate the traffic likely to be generated by a development.

Some planning permissions for major developments will require an environmental impact assessment to be carried out prior to the granting of planning permission. A guide to the types of developments affected can be obtained from the Department of the Environment's Circular 15/88 Town and Country Planning (Assessment of Environmental Effects). Some very significant developments will always require environmental assessment. These are known as Schedule 1 Projects and include such things as motorways, major airports, chemical installations, heavy industry and thermal power stations. Other developments, known as Schedule 2 Projects, require environmental assessment when they are likely to give rise to significant environmental effects and annex A of Circular 15/88 contains some indicative criteria and thresholds for identification of such projects requiring environmental assessment. For example, an industrial estate development may require environmental assessment if the area is in excess of 20 hectares. The preparation and submission of an environmental impact assessment will usually involve consideration of a wide variety of impacts which the proposed development might have on the environment, consultation with a number of statutory bodies and other organisations, compliance with publicity requirements and, ultimately, producing a written document reporting environmental impact in the form of an environmental statement. Inevitably, the production of the environmental statement is time-consuming and if environmental assessment is required, then the time allowed for compliance should be considered in preparing a timetable for the proposed project.

Other Permissions

A number of other permissions may be required in connection with the development of an urban regeneration scheme or project.

Listed Buildings Consent

If the development involves works which will affect the character or setting of a listed building, then in addition to a grant of planning permission, it will be necessary to obtain listed building consent. Details of whether or not a building is listed can be obtained from the local authority through a search; however, it should be noted that a search in relation to the immediate site may be insufficient in this respect as the development may affect the setting of listed buildings which are on the edge of but outside of the site itself. In this case, listed building consent may still be required. It is important to make an early assessment of whether listed building consent will be required, because the procedure for obtaining listed building consent is very similar to the procedure for obtaining planning permission and the two often run hand in hand, and are subject to similar time frames and rights of appeal.

Tree Preservation Order Consent

Trees may be protected by a Tree Preservation Order. Such an order will be revealed by a local authority search. If it is intended to cut down or otherwise damage trees protected by a Tree Preservation Order the consent of the local planning authority must be obtained.

Conservation Area Consent

A local authority search will reveal whether or not a site lies within a conservation area. If it does, then conservation area consent may be required for some aspects of the development. Again, the procedure is very similar to the planning application procedure. In any event, any application for planning permission will be considered more stringently in terms of the proposed design of the scheme or project if the site lies within a conservation area. Trees in a conservation area are also subject to special protection.

Designated Areas of Archaeological Importance

If any part of a site is designated as an area of archaeological importance, compulsory archaeological research facilities will have to be provided to archaeologists prior to redevelopment of the site. This can have implications in terms of the time within which a regeneration scheme can be completed, because there must be a mandatory delay of four months and a period of two weeks for archaeological research before any operations are

carried out which disturb the ground or where flooding or tipping operations are proposed. Even where a site is not so designated, a local planning authority is entitled to attach conditions to any grant of planning
permission requiring archaeological investigations to be carried out if there
is any evidence that the redevelopment may affect archaeological remains.

Some Special Areas

In certain designated areas, special conditions and circumstances are in
force that alter the normal operation of planning law. These include Enterprise Zones and Simplified Planning Zones.

Enterprise Zones

An enterprise zone is an area designated by the Secretary of State with the
object of stimulating industrial and commercial activity by giving substantial financial advantages to developers within the zone. Planning controls in
an enterprise zone are simplified. The reduction of planning controls in the
zone areas depends primarily on the contents of the document known as
the 'Enterprise Zone Scheme' which is prepared by a zone authority in
advance of the designation by the Secretary of State. From this it will be
possible to deduce which planning permissions are automatically granted in
the zone and which conditions or limitations (if any) apply.

Simplified Planning Zones

The purpose of a SPZ is to allow a local planning authority to grant a
general planning permission for some part of its area. Within such an area,
the SPZ developers are able to undertake developments as of right, up to
the tolerances specified by the SPZ scheme, and without requiring further
planning permission. Simplified Planning Zones are an extension of the
Enterprise Zone concept, although without the fiscal advantages of EZs.
Schemes for SPZs are prepared by a local planning authority, following
publicity and consultations. There is a right to object to a draft scheme,
and, a public inquiry must be held to consider any objections which are not
met by the local planning authority. Once an SPZ is adopted, a copy of the
scheme can be obtained from the local planning authority in order to
ascertain the types of development which can be carried out within it,
without the need for a further grant of planning permission.

Conclusions

This chapter has provided on overview of some of the major areas of law
that are relevant to the preparation and implementation of an urban regeneration scheme or project. As can be gathered from the text, there are
many individual circumstances and situations in which special conditions or

---

Key Issues and Action

- Consider structure at the outset – is a 'special purpose company' appropriate?
- Involve relevant agencies (e.g. new Regional Development Agencies).
- Consider regeneration schemes with local authorities and central government encouragement for PFI schemes and public/private partnerships.
- Note continued relaxations to capital finance regime relating to disposal of asset by local authorities.
- Identify the interests that need to be acquired to assemble the site.
- Identify the rights and covenants on the site which may impact on the proposed development.
- Work out a strategy for securing the necessary interests in the site without committing the developer to purchase the site before he is ready to proceed with the development.
- Carry out an environmental investigation of the site and take account of the results in determining the strategy for purchase, development and sale on.
- Identify what planning and other permissions are required for the development, the likelihood of obtaining them and the relevant procedure for application.
- Identify any possibilities for delay in the planning process, e.g. referrals to the Secretary of State, and ensure they are taken account of in the development strategy.
- Ensure compliance with waste management laws during the development phase.

---

aspects of the law apply. However, in general, the key message of this chapter is that it is essential to obtain good legal advice at the outset of an urban regeneration scheme. By adopting this approach, many potential difficulties and obstacles can be identified in advance, and the appropriate permissions or agreements can be obtained or achieved in good time, thereby avoiding any unnecessary delays or excessive costs. Anticipation and early action can prevent difficulties from becoming problems.

There are many aspects of the law that have not been discussed in this chapter; the material that has been presented is considered to be relevant to most urban regeneration activities: in particular it is essential to agree and implement appropriate structures that will enable the regeneration scheme to proceed in an effective and efficient manner.

Further Reading

DETR: (1997, 1998) *Local Government and The Private Finance Initiative – an Explanatory Note on PFI and Public/Private Partnerships in Local Government*, DETR, London.

Grant, M. (1882) *Urban Planning Law (supplemented 1990)*, Sweet and Maxwell, London.

Hellawell, T. (1995) *The Law Society's Environmental Law Handbook*, The Law Society, London.

Megarry, R. and Wade, W. (1997) *The Law of Real Property*, Sweet and Maxwell, London.

10 Monitoring and Evaluation

Barry Moore and Rod Spires

Introduction

Measuring, monitoring and evaluating urban regeneration is a vital task. Indeed, the availability of financial and other forms of support for projects and programmes is normally tied to the provision of an acceptable framework for monitoring and evaluation. In addition, given the wide range of actors and organisations involved in urban regeneration, it is important to be able to demonstrate the outputs of initiatives and to be able to point to the origins and consequences of any difficulties that have been encountered during the process of implementation. In broad terms, monitoring and evaluation attempt to identify what actions have taken place and what the consequences of such actions have been.

This chapter presents a number of issues:

- the general principles of monitoring and evaluation;
- the importance of designing strategies in order to allow for the incorporation of monitoring and evaluation;
- the measurement and monitoring of progress; and
- the evaluation of urban regeneration strategies.

At the outset is it important to recognise that the task of monitoring and evaluation is closely linked with policy development both at a strategic level and when specific projects are being designed and implemented. As such it forms part of the policy process and is related to policy choices and the establishment of aims and objectives. These choices can be influenced by political aims which in turn set the context for monitoring and evaluation activities. Hence the approach to evaluation, the choice of what is measured, and the judgement as to what has been achieved cannot be divorced from the wider political or cultural context.

A related issue is the desirability for the evaluation task to be seen as reasonably objective in that it does not rely solely on the views, and judgements, of those directly involved in policy formulation and implementation. In this sense impartial advice usually forms part of the evaluation process.

The nature of evaluation is also influenced, among other things, by resource availability in terms of skills, personnel (in-house or external) and capacity to collect, organise and analyse information, including data.

Resource availability will determine the breadth and depth of the evaluation task.

Timing is also a critical issue. At the early stages of policy implementation the emphasis is more likely to be on monitoring actions. As initiatives
mature the focus will be on outputs, outcomes and added value either as
part of interim or final evaluations. At this stage issues of effectiveness and
efficiency become more important.

Principles of Monitoring and Evaluation

Measurement, monitoring and evaluation can be considered as an integral
part of the cycle of urban regeneration. The cycle starts with the identification of the challenges to be addressed; it continues through the various
processes of planning and strategy; it then progresses to the point of implementation; and, eventually, to completion. At all of these stages in the cycle
it is important to be able to:

- draw upon the experiences of previous projects and programmes in
  order to help to identify and avoid problems and the potential waste of
  resources;
- to identify targets and to incorporate them within an agreed schedule of
  action and implementation;
- to measure and monitor specific aspects of implementation;
- to evaluate the overall performance of a project or programme, that is,
  effectiveness and efficiency.

It is helpful to address the question of how best to establish methods of
measurement, monitoring and evaluation at the start of the urban regeneration cycle, and to incorporate the necessary measures and measurement procedures within the plan or strategy. Many of these procedures and
practices associated with monitoring and evaluation are familiar to urban
regeneration practitioners. Individuals and organisations in both the public
and private sectors are expected to monitor and evaluate actions as part of
their routine activities, whilst community and voluntary groups are normally keen to ensure that the best use is made of their limited resources.

Initially it is important to attempt to clarify two aspects of measurement,
monitoring and evaluation:

- the purposes of the exercise;
- the terminology used, especially by European Union and central government policy-makers.

The Purposes of Monitoring and Evaluation

The purposes of measurement, monitoring and evaluation have already
been hinted at in the preceding paragraphs. Evaluation is a key tool in

gauging the extent to which policies and initiatives are effective and efficient in terms of meeting aims. Evaluation provides a basis for judging whether there is still a rationale for policy intervention (or policy needs to be adjusted) and whether implementation is resulting in the designed outcomes in the required time-scale. The specific aims are: to check the progress of a project or programme against specified targets in a systematic and transparent manner, to inform the review or revision of the original targets and actions, and to arrive at a judgement overall of the outputs of the scheme and the added value (or additionality) it brings. Each of these purposes is outlined and explained in the following sections. A more extensive discussion of monitoring and evaluation is provided in the later sections.

Progress

Any urban regeneration project or programme is designed with the intention of achieving a number of specific requirements and targets. These requirements and targets, normally used in order to help design a project or programme, are usually incorporated in the development and business plans, and are also presented as part of the case which is put forward for funding. A typical set of targets will cover a variety of initiatives and will identify the expected outputs and units of measurement (or indicators) that are appropriate to the activity in question. Having specified and agreed the targets it is then necessary to put into place procedures to check and report progress sometimes against a baseline position; these procedures may include collecting management information, land-use surveys, other direct surveys (of, for example, firms or occupants) and the requirement that the recipients of funding should report their achievements. Indirect monitoring of other sources of information such as employment creation and the local level of unemployment may also be undertaken. By setting targets and then measuring and monitoring progress, it is possible at a given point in time to assess the extent to which the initial objectives have been achieved. An example is provided in Table 10.1 of the level of achievement of the Trafford Park Development. The level of achievement reached by March 1995 is compared with the level of performance specified in the Corporate Plan.

Revision

An additional purpose of measuring and monitoring progress is to assist in the review and revision of a plan or strategy and the identification of any new challenges which may emerge. For example, in the case of the example provided in Table 10.1, if an unexpected industrial closure occurred, then the Trafford Park Development Corporation would find it necessary to revise its expected targets and, in such circumstances, the corporation may wish to introduce additional measures in order to attract additional companies to locate in the park, to encourage existing companies to expand their activities, or it may need to revise the entire strategy. Such review and revision is normal and should not be interpreted as indicating the failure of

Table 10.1   **Progress achieved by Trafford Park Development**
              Corporation

	Target level	Achievement by March 1995
Land reclaimed	170 ha	142 ha
Highways built or upgraded	40km	31km
Commercial development	600,000m²	496,000m²
Training places assisted	3,500	2,420
Gross jobs created	19,000	16,200
Companies attracted	800	720
Private sector investment	£1,200 million	£915 million
Trees and shrubs planted	800,000	759,000

Source: Trafford Park Development Corporation (1995).

either the original strategy, or the process of implementation. However, it may be the case that the original target was unrealistic or that unexpected problems have emerged which could have been anticipated at the outset. Hence the evaluation process can be used as an input to policy adjustment or to keep policy on track.

Judgement Overall

At the end of an urban regeneration project or programme it is essential to evaluate the overall level of performance and the causes and consequences of any significant shortfalls or overshoots compared with the targets and goals specified at the outset. Identifying the causes and consequences of any variations is important because the examination of unexpected successes or failures may help to reveal or demonstrate:

- ways of working to be avoided or encouraged in future;
- the best way of addressing a given problem;
- the ever-present influence of external events;
- the likelihood that the unexpected will occur.

Terminology

The terminology that is employed in discussions of measurement, monitoring and evaluation can be open to interpretation. In most cases the ideas and procedures that are associated with monitoring and evaluation are relatively straightforward and can be explained and applied without generating any undue uncertainty. However, the excessive use of jargon and terminology can detract from these aims and can reduce the value of the outputs of evaluation for practitioners and policy-makers.

Notwithstanding the observations made in the preceding paragraphs, it is essential to agree the basic terminology and definitions that are in general use. Although specific terms are frequently employed by individual initiatives or organisations, the most commonly used terms include those listed in the Glossary at the end of this chapter. These definitions have been drawn from a number of UK Government and European Union documents including *Single Regeneration Budget Bidding Guidance* (Department of the Environment, 1995) and *Policy Evaluation: A Guide for Managers* (HM Treasury, 1988). A knowledge of these terms will be helpful in reading the remainder of this chapter.

Developing the Regeneration Strategy

Measurement, monitoring and evaluation commence early in the process of urban regeneration. The very act of defining the problem requires existing, or baseline, conditions in an area to be measured and to be compared with the local, regional or national coverage. Initially attention is devoted to defining the criteria used to determine the eligibility of an area or activity to receive support, and an equal or greater amount of time and money is spent by sponsors in seeking to demonstrate that their project or programme meets the eligibility criteria.

This section of the chapter sets out some of the key stages in developing an urban regeneration strategy. Although the topic of strategy has been discussed in more detail in Chapter 3, the purpose of including this discussion here is to demonstrate that the process of strategy formulation cannot be separated from the measurement, monitoring and evaluation of programmes and projects. The strategy development process breaks down into several well-defined stages:

- identifying the scale, nature and causes of urban problems;
- reviewing current policies and programmes;
- setting strategy objectives;
- appraising strengths, weaknesses, opportunities and threats.

Each of these stages is now considered below in turn, although in practice they may overlap.

Identifying the Scale, Nature and Causes of Urban Problems

The current framework of analysis within which urban regeneration strategy development occurs, and in particular the diagnosis of the problems to be addressed and their causes, places great emphasis on the proper functioning of competitive markets as the means by which resources such as unemployed labour or vacant and derelict land can be efficiently utilised.

A key argument for promoting economic regeneration is that markets are not working properly. This may be because of institutional constraints that prevent markets from working freely (or adjusting quickly), or an external effect that is

---

Box 10.1   An example of market failure: the urban land market

Failure in the market for land occurs when developers' profits are eliminated as a result of excessive asking prices for sites, leading to an undersupply of land at prices that can be supported by an assessment of residual value. The source of the problem lies in information costs and the landowner's perception of risk. Thus developers have to be prepared to invest time and money in researching the current value of a site and successful projects carry the abortive work on projects not secured or that do not go ahead. By contrast the landowner is unlikely to devote similar resources in establishing the current value of the site and neither will an agent acting on behalf of the landowner. The costs to the landowner of making the wrong decision either to hold or to sell are asymmetric. A wrong decision to hold loses the value of that offer and the landowner will continue to bear the holding costs. However, these costs are likely to be very small by comparison with the costs of not holding on to land which subsequently increases in value because of, say, a change in planning permission. Thus the average expected value of sites to a landowner is likely to exceed the average current offer and the supply of land is restricted. The *raison d'être* for public policy intervention is that it provides a mechanism for the sharing of risk and information with the landowner, thereby facilitating the decision as to which land to hold and which to sell and improving the efficiency of the land market.

---

not properly reflected in market prices. Correcting market failures improves the supply side and increases the productive potential of the economy as a whole, either in the short term (by promoting flexibility and more rapid adjustment of the economy to external shocks) or in the longer term (by increasing productive capacity).   (HM Treasury, 1995, p. 4)

Within this traditional model the economic case for government and other intervention is based primarily on evidence that markets are failing to work properly and the role of policy is to correct such market failure or compensate for it. The causes of market failure may be a consequence of both public and private sector decision-making and this should be recognised in approaching this first stage in developing a strategy. Although the above discussion focuses on the urban regeneration problem from an economic efficiency perspective, it is of course the case that many of the problems associated with cities are social problems in which the issues are essentially distributional and political. An example of market failure – the case of the urban land market – is illustrated in Box 10.1.

The urban problem is also a complex problem and generally is the outcome of a set of interrelated problems which combine to reduce the competitiveness of firms in a town or city, the competitiveness of individuals in the labour market and the competitiveness of the area in attracting inward investment of physical, financial and human capital. An inner city area for example, may be suffering from market failure in several markets – the labour market, the financial and capital markets, the property market and in the upgrading and provision of infrastructure – with market failure in one area interacting with and potentially reinforcing market failure in another area. Thus market failure in the commercial property market may give rise to a dearth of suitable accommodation for inward investment or new firms. This in turn may reduce the prospects of the local unemployed finding employment, thereby exacerbating market failure in the local labour market as the duration of unemployment increases and the skills of the unemployed labour force wither. In these circumstances, the strategy must recognise the need for an interrelated programme of initiatives, addressing a range of problems and with policy priorities being determined by the perceived relative importance of different problems.

Not only is the problem to be targeted complex, it is also an evolving problem, and a difficult issue to be faced in designing a strategy is that of identifying the scale and nature of the problem and its likely future trajectory. For example, although the current rate of inner city registered unemployment may at first sight appear to provide a useful indicator of the extent of market failure in the local labour market, it provides only limited guidance for the policy-maker, because:

- it provides no information on the nature of inner city unemployment such as the average duration of unemployment, the age and skill structure of the unemployed and its geographical density;
- it will not include those who are seeking work but not registered; it provides no indication of whether the problem is likely to worsen significantly in the future.

An important aspect of problem definition is that of specifying target groups and target geographical areas. For policy to maximise cost-effectiveness it is essential that the outputs of policy reach those groups for which it is designed. For example, a policy targeted on creating jobs for inner city unemployed residents may fall short of achieving its objectives if commuters from outside the inner city area successfully compete for any jobs created in the inner city and if any jobs thus made available elsewhere are inaccessible to inner city residents. Another example is that of subsidies paid to firms, which may partly end up in profits with perhaps only limited benefits to the local urban economy and its residents. These examples illustrate the dangers of policy missing the target if there is no impact model guiding those responsible for developing policy.

Considerations of cost-effectiveness also raise questions concerning the responsiveness of different groups to policy intervention. A training programme to improve the competitiveness of different groups in the local labour market may be much more effective for those only recently made unemployed by comparison with the impact on the long-term unemployed.

Although a policy programme may be well targeted, participation by potential beneficiaries may not be mandatory and how a programme is presented, organised and delivered will affect the take-up of the policy. In this respect the outputs of the policy may depend critically on the policy delivery system.

Review of Current Policies, Programmes and Partnerships

An essential stage in the development of an urban regeneration strategy is a review of current programmes and projects, including the identification of the main organisations/partners involved and the mechanisms and institutions established to deliver the policy and secure the strategic objectives currently in place. It is also important to establish what is *not* being done and where there might be a need to refocus the strategy. An important part of the review will also be concerned with auditing the levels and sources of funding and its allocation across programmes and projects.

Setting Strategy Objectives

The objectives chosen for the strategy should relate explicitly to the problems identified and to their underlying causes and, most importantly, to the agreed priorities of the strategy. They should also be realistic and achievable, and recognise the constraints within which the strategy is formulated. This suggests that there is no simple definition of the objectives of policy intervention aimed at regenerating urban areas, but rather a hierarchy of objectives at the top of which is an agreed overall aim of policy followed by the strategic objectives of a particular programme or project. Having defined the specific objectives of a programme or project it is possible to identify specific operational objectives and objectives associated with individual targets and milestones.

For example, in the 'Inner Cities Initiative Fact Sheet' there was a statement on the ministerially agreed overall aim of the project, followed by a list of four strategic objectives and a further list of six secondary objectives. Thus the overall aim of the Initiative was 'to improve the targeting and enhance the benefit to local people of the money channelled through existing central programmes'.

The four main strategic objectives were:

(1)   To provide more jobs for local people by removing impediments to their recruitment by local firms and encouraging local enterprise development.
(2)   To facilitate enterprise by local people through enterprise training and financial and managerial assistance.
(3)   To improve the employability of local people, including those newly entering the labour market, by training programmes aimed at specific employment opportunities or gaps in the labour market.
(4)   To support projects designed to improve the environment, and the provision of community services and to reduce the level of crime where these can be linked to the re-integration of local people into local economic activity.

In addition, there were six secondary objectives:

(1) To stimulate economic activity and employment by using task force money to pump-prime private sector involvement and investment in the area, to lead to a firmer basis for further economic development.

(2) To improve the co-ordination between different government programmes and with the activities of other local bodies such as local authorities, the voluntary sector and private industry.

(3) To sensitise government departments to employment and enterprise issues in inner cities, to identify impediments to the operation of departmental programmes and, where possible, to adapt programmes to make them more accessible to local people and more relevant to their needs.

(4) To strengthen the capability of local organisations to undertake long-term economic enterprise development.

(5) To target the employment needs of specific disadvantaged groups especially ethnic minorities.

(6) To develop innovative approaches to problems which are capable of application in other inner city areas

The current SRB programme also has a very wide-ranging set of objectives and the *Bidding Guidance* indicates seven key areas covering economic, physical and social conditions. These include the improvement of employment opportunities for local people through training, improved competitiveness of local firms, physical and environmental improvement, better housing for local disadvantaged groups, a better quality of life for local residents, reductions in crime and improved community cohesion and improved access for ethnic minorities.

Appraising Strengths, Weaknesses, Opportunities and Threats

Figure 10.1 provides a framework within which the SWOT analysis may be carried out. Five main external drivers of change are identified – economic, demographic, social, technological and public policy. These drivers of change give rise to threats and opportunities which affect the competitive position of the city, including its economic performance, labour market performance, cohesiveness and sustainability. For example, public policy changes emanating from the European Union or the UK government will influence the competitive position of different firms and sectors operating in a city and its subregion, the effectiveness of the labour market and the sustainability of urban development. European and global economic trends will also impact on the city's socio-economic development as different sectors adapt to the changing competitive environment within which they operate.

The effectiveness, speed and capacity for adaptation to the changing external environment, including competition from other cities and regions, reflect the city's strengths and weaknesses and these are captured in the bottom half of Figure 10.1. They range from the inherited structure of

Figure 10.1. **A Framework for Undertaking a SWOT Analysis**

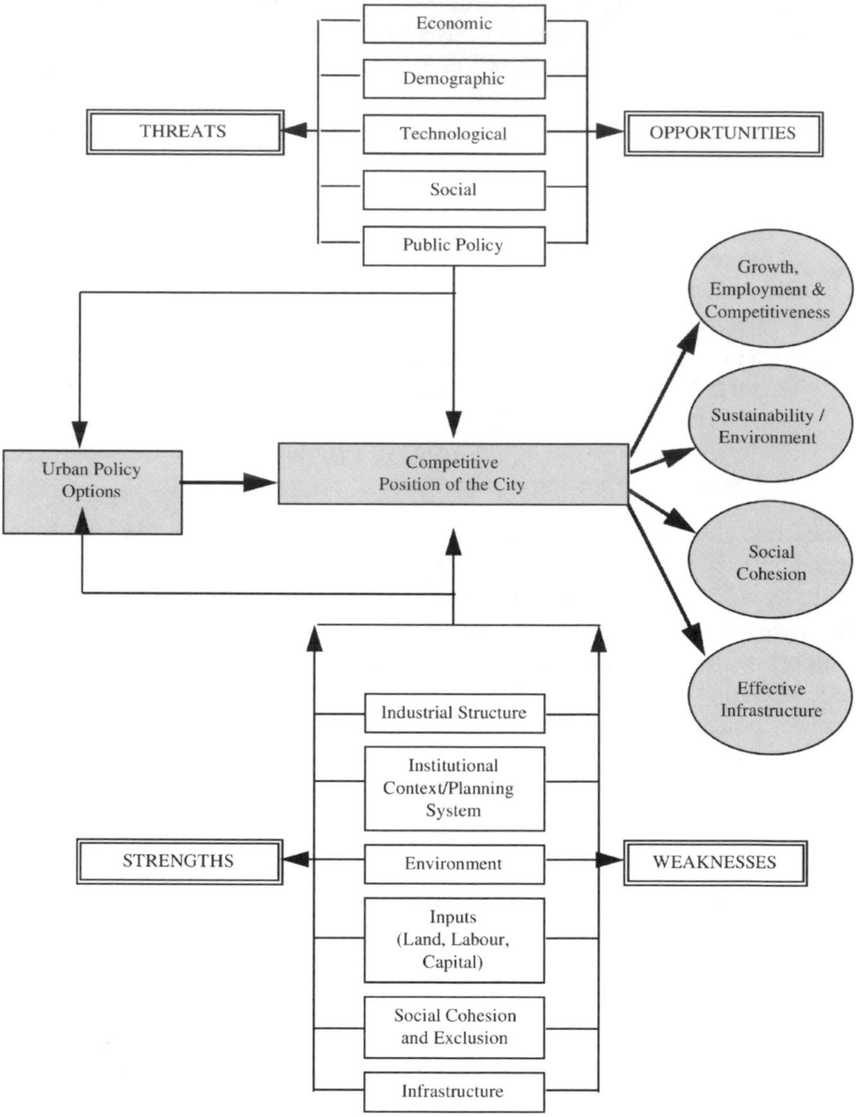

Source: PACEC (1999)

industry and economic activity, the institutional and policy structure, the availability, quality and cost of productive inputs, the business and social infrastructure and the degree of social cohesion and exclusion. These inherent strengths and weaknesses condition the extent to which threats are effectively resisted or overcome and opportunities exploited. Policy acts to mediate and influence this interaction in a number of ways; by correcting market failure; supporting the development and enhancement of the infrastructure; ensuring that outcomes are sustainable and environmentally acceptable; and by securing an acceptable social (and spatial) distribution of benefits and opportunities. The right-hand side of Figure 10.1 shows the broad outcomes of these interactions for the development of the city and its subregion which, although shown separately, interact in a variety of ways.

Measuring and Monitoring Progress

This section of the chapter is concerned with the measurement and monitoring of urban regeneration policies, programmes and projects. Three different roles are generally associated with the monitoring of programmes and projects:

- monitoring of the conduct of a programme or project including issues of eligibility, compliance, programme or project coverage and the identification of recipients of expenditure grants and other forms of support to ensure that the programme or project is being properly conducted within the agreed institutional and legislative framework;
- monitoring of progress in achieving intermediate objectives relating to programme or project co-ordination, the bending of public expenditure in support of regeneration objectives, responses of the private/ voluntary sector, the leverage exerted and matters concerning the programme or project;
- monitoring indicators which relate to the key objectives; these derive directly from the problems or issues faced by an area and where improvements are sought.

Area Targeting

Because many aspects of urban regeneration programmes are highly targeted on specific small geographical areas and on particular disadvantaged groups, establishing programme or project coverage is a critically important task for monitoring activities. It is primarily concerned with estimating the extent to which the programme reaches its intended target population. Closely linked to this is the monitoring of programme or project delivery. This assesses the extent to which planned activities and targets correspond to the actual way intervention occurs and how the delivery may be improved.

Monitoring and Management

The monitoring of programmes and policies is driven very much by the information needs of managers and is often a component of management information systems. However, evaluations also make substantial use of monitoring information and, indeed, the dividing line between monitoring and evaluation is often somewhat blurred although the former is more concerned with actions and take-up rather than impact. In addition, policy-makers and other stakeholders in the policy process have information needs to be met. From the evaluator's perspective monitoring is important for understanding and interpreting evaluation findings. Monitoring information is also very helpful where the evaluation requires a sample survey to be undertaken, by providing a useful sampling frame for the sample selection. Monitoring information may also be helpful in establishing suitable control or comparison groups. Evaluators also need information on the extent and ways in which individual programme elements are delivered and this is particularly important where, as in SRB or European Structural Fund programmes, partnership is a central part of the policy process. Monitoring information is also critical from the perspective of those who fund and support programmes and projects by providing information on the activities undertaken, the extent of implementation, the take-up of programmes, and whether eligibility criteria are being properly adhered to. The provision of reliable financial information is another valuable input into the evaluation process that results from the monitoring process.

Participation and Take-up

Measuring and monitoring the extent to which a programme or project reaches its intended target is essential particularly in economic and social programmes in which participation is voluntary, such as in some urban community programmes or business support initiatives. Managing a project or programme efficiently requires accurate and timely information on target participation and this is particularly so with innovative schemes where adjustments may be necessary if take-up falls seriously below target. A bias in participation and coverage may occur partly as a consequence of self-selection, differences in information access by different groups and because of rejection or reluctance to participate in the programme.

Effective Delivery of Programmes and Projects

The final area for monitoring concerns the delivery of an intervention or action, for often the failure to show real impacts may be a failure to deliver the appropriate type of service or support and this is a charge often levelled at educational and training programmes targeted on disadvantaged groups. Delivery systems may also fail because of inadequate structural and organisational arrangements which impede access to a policy, programme or project.

The emergence of interim evaluations, which are undertaken relatively early on after the start of a programme or project, reflects the increasing importance attached to the development of intermediate output. Such indicators or monitoring measures include indicators of, for example, the physical progress made with respect to environmental improvement or land reclamation, throughput of trainees on training programmes, number of businesses supported, inward investment projects attracted, number of dwellings improved, number of community schemes introduced or recreational facilities provided. These and many other indicators of intermediate outputs are of value not only in providing invaluable monitoring information on the progress of a project or programme but also as the starting-point for the evaluation exercise itself. Those indicated at the early stages may be both quantitative (management information) and qualitative (soft information).

Evaluation of Urban Regeneration Strategies

The evaluation of urban regeneration strategies is a complex business and differs from measurement and monitoring in a number of important respects:

- There are obvious technical and conceptual difficulties relating to the theoretical framework to be used for understanding how different policy initiatives influence private and public sector decision-making and the cause–effect relationships that need to be articulated if the policy effects are to be satisfactorily disentangled. Evaluations usually need to recognise that the institutional framework within which policy is delivered will influence both how policy works and with what effect, suggesting that similar policies may differ considerably in their impact across different cities and local areas.
- Evaluation is concerned with the added value of initiatives and the net additional outcomes which contribute towards aims and objectives, and help to correct market failure. Effectiveness and efficiency are interrelated issues in evaluation.
- There are practical difficulties, and choices, concerning the choice of suitable indicators for measuring the effects of policy, the collection of data necessary for constructing indicators, the timing of policy evaluations, the relevant spatial unit of analysis.
- There are analytical and statistical difficulties of how best to disentangle the impact of projects and programmes.

What is clear is that there is no agreed consensus as to how these difficulties might best be met and the research that would be required for this purpose. To a very significant extent evaluation research is being driven by political pressures designed to ensure a degree of accountability, comparability between projects and programmes in terms of their achievements, and a realisation of the potential for transferability of policies and good practice between urban areas in different regions and countries. Arguably it is in the context of the latter that an accepted evaluation framework has emerged during the 1980s and 1990s and such a framework is perhaps best captured in HM Treasury's guidance document published in 1995.

Numerous applications of this evaluation framework may be found in a wide variety of studies commissioned by various government departments, the European Commission and economic development agencies. This section aims to provide an overview of what is here termed the 'Treasury Evaluation Framework' (TEF) although it is accepted that in certain respects it may depart from the guidelines and framework set down by the Treasury in 1995.

Within the TEF the primary purpose of evaluation is to assess the degree to which policy, programme and project objectives are being secured, and how effectively, efficiently and economically they are being achieved.

Efficiency

The efficiency of a policy, programme or project relates outputs to resource inputs and establishes what might be called a measure of value for money (VFM). Such a measure of performance may be carried out with respect to the strategic objectives of the policy, the programme as a whole or with respect to specific projects.

Developing this performance measure clearly raises a number of conceptual and measurement problems:

- Not all costs and outputs (benefits) associated with urban regeneration can be expressed in monetary terms in order to permit the development of a simple efficiency indicator relating inputs to outputs, e.g. improved confidence, reduced fear of crime, improved community spirit, increased local commitment and capacity-building.
- Intermediate outputs associated with particular projects or activities such as training throughput, number of hectares of land reclaimed, factory floor-space provided, small businesses assisted, may be misleading in establishing the final outputs or benefits of policy which typically relate to employment and value added.
- Assigning final outputs to specific activities such as training, business advice or property provision may not be possible because each may contribute jointly to employment or value-added benefits.
- Typically the operation of an urban regeneration scheme is associated with a wide range of outputs some of which may be measured only qualitatively and this may preclude the possibility of developing a single comprehensive measure of a policy's outputs. For example the SRB is concerned with securing a mix of economic, social and physical objectives and, whilst some of the benefits of this programme will be quantifiable, a number will inevitably be qualitative in character.
- Outputs and inputs accrue through time and this raises the question of when to measure the different inputs and outputs, whether the policy outputs associated with a particular input have fully matured and whether to discount public expenditure flows and output flows. Moreover some policy outputs will remain after policy has been

withdrawn, whilst in other cases outputs require a persistent policy injection if they are to be maintained.

- Some policies such as local planning or zoning controls have limited cost associated with them in terms of public expenditure but may give rise to significant opportunity costs if the productivity of business is adversely affected, and this may be very difficult if not impossible to measure.
- Where incremental changes are being made to a programme, marginal not average costs and benefits should be used in deciding on the policy adjustment.

Effectiveness

A second policy performance measure is that of effectiveness. This measure compares the outputs of policy with the output objectives of policy. However, as discussed above, the objectives of urban policies and regeneration strategies are rarely unambiguously set down or stated in a quantitative form. Moreover, given that the paramount rationale for policy intervention within the TEF is market failure, central to the measurement of effectiveness must be an assessment of the extent to which market failure has been corrected, moderated or, indeed, worsened.

Economy

Economy is a measure of the actual resource inputs or costs of a policy relative to the planned or budgeted costs. This is important when the take-up or participation in a policy, programme or project is uncertain and it may also be a helpful indicator of possible inefficiencies and wasteful expenditures.

Other Indicators

In addition to the three core policy performance measures – efficiency, effectiveness and economy – a number of other indicators have been developed by those concerned with evaluating policy. They relate to issues such as targeting of policy, sufficiency of policy to meet the scale of the problem and the acceptability of policy on the part of policy consumers. The following summarises some of these additional performance measures:

- Leverage is a measure of the extent to which public expenditure on a particular programme or policy attracts financial support and other resource inputs from the private sector. It is becoming an increasingly important indicator as pressures on public spending persist and grow, and as new partnerships between public and private sector organisations develop. A commonly used measure of leverage is the ratio of total inputs to policy inputs.

- Targeting is a measure of the extent to which the outputs (or benefits) of policy are secured or enjoyed by the intended beneficiaries and may be measured by the proportion of policy outputs enjoyed by the target group. Where the incidence of a policy input may be shifted the concept may also be used on the cost as well as the benefit side of the policy balance sheet.
- Sufficiency assesses the extent to which the outputs of policy resolve the underlying cause of the problem, be it one of market failure or distribution.
- Acceptability compares the policy, programme or project being delivered with the policy that consumers prefer.

The performance measures identified above refer primarily to outputs and benefits relating to specific urban or geographic areas, but it is important to realise that there are circumstances whereby national benefits may arise. In the absence of these types of impact urban regeneration policies serve mainly to displace economic activity and secure distributional benefits.

A Framework for Evaluation

In order to assess the performance of policy using the different indicators set out above, it is necessary to evaluate the different benefits arising as a consequence of policy, the beneficiaries of the policy and the financial and other costs associated with the policy. A methodological framework for achieving this is set out diagrammatically in Figure 10.2. The framework reflects current Treasury guidelines and current urban policy evaluation practices. It distinguishes:

- strategic objectives;
- inputs and expenditures;
- activity measures;
- output and outcome measures;
- gross impact measures
- net impacts.

The top of Figure 10.2 identifies the broad strategic objectives of urban regeneration programmes including SRB, and embraces economic, physical and social objectives. Each of these broad areas will be associated with a particular set of strategic objectives made up of activities reflecting the priorities and needs of the local area. The activities will be measured in different ways and examples are provided in the 'Activity Measures' boxes. The activities will in turn generate outputs and outcomes that can be measured net and gross, which give rise to the ultimate impacts on the local economy, for example, jobs, wealth, capacity-building. There is a causal relationship between these elements. The entries in the boxes of the Figure 10.2 flow diagram are illustrative only.

Figure 10.2: **Ex post Evaluation Framework for Urban Regeneration Policies**

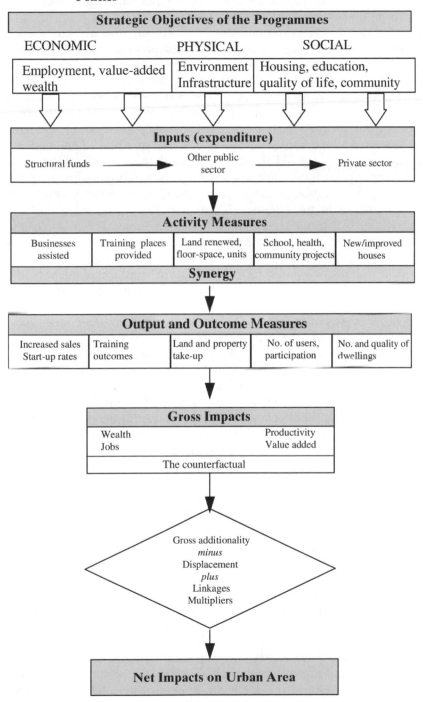

Source: PACEC (1999)

Central to the evaluation framework and the empirical work designed to evaluate policy is the need to distinguish gross and net outputs and this is indicated in the bottom half of Figure 10.2.

At issue here is the vexed problem of identifying and measuring what otherwise would have happened if a policy or programme had not been pursued i.e. the counterfactual. Establishing effectiveness amounts to establishing cause and effect, and additionality arises when there is a departure from what otherwise would have happened had the policy intervention not occurred, that is, there is a departure from the counter-factual. Thus additionality exists when the outputs or benefits of the policy would not have been generated in the absence of the policy. The question of additionality can arise both with intermediate outputs and with final outputs of policy.

The starting-point for establishing additionality is to distinguish gross outputs from gross additional outputs of policy. Gross outputs consist of all observed changes in the output indicator over the period when policy is being evaluated. Thus the gross output of policy measure designed to create more jobs in a particular city might be the change in the number of jobs in the city in the period of policy intervention. The gross output of a project concerned with providing new factory floor-space in a locality might be the change in new factory floorspace associated with the policy project concerned. In both of the above cases it could, however, be argued that, in principle, all or part of the change in jobs or factory floor-space might have arisen in the absence of the policy intervention. In the case of the property market it is conceivable that the policy could discourage some provision by the private sector that might otherwise have occurred. Equally there may have been adjustments in the labour market in the absence of policy. The measure of what otherwise would have happened to the selected output indicator in the absence of policy intervention is termed the dead-weight.

To establish the net additional impact of policy gross additional outputs must take account of substitution, displacement and any indirect impacts such as input/output linkages and multiplier effects. Substitution occurs on the supply side and arises when policy-targeted resource inputs take the place of other inputs, for example, labour trained as a result of a local training initiative takes the place of labour that would have been employed but has not been through the policy initiative. Displacement occurs when the extra output from a policy-supported activity leads to less output from firms not directly participating in the policy initiative – for example, a subsidy paid to a particular firm in an industry improves the competitiveness of that firm such that other firms in the same industry lose market share and reduce their output and employment in the local area. Finally indirect effects arise as a result of the additional economic activity in the local area. Thus increased output by a firm may give rise to an increased demand for locally produced inputs. Increased employment gives rise to increased wage and profit income, part of which will be spent in the local economy thereby generating further output, income and employment. These concepts are illustrated in Figure 10.3.

Figure 10.3 **Distinguishing Gross and Net Policy Impacts**

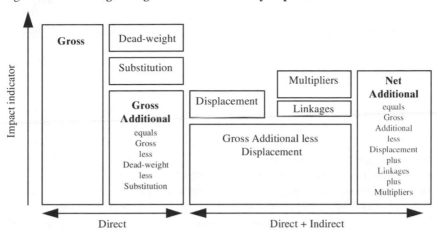

Source: PACEC (1999)

The strength of this evaluation framework is that programme impacts are clearly linked back to its objectives and inputs and key intervention categories are identified. At the same time it recognises the importance of measuring and distinguishing activity and output measures in the achievement of ultimate impacts. It also focuses on the need to distinguish gross from net effects at each stage in the evaluation process – ex ante, intermediate and ex post.

Programme or Project Costs and Inputs

Identifying and measuring the costs associated with an urban regeneration strategy is an important step in developing performance indicators for judging whether the programme or project is worthwhile, whether more resources should be devoted to it and how resources allocated to different projects within the programme might be modified to improve overall effectiveness. Policy-makers, managers and funders are therefore inevitably faced with the problem of how best to allocate scarce resources to achieve the objectives of the regeneration strategy.

When estimating the costs of a programme both public and private sector costs can be included. In the case of the public sector, both directly associated programme costs as well as any other public expenditure necessary to secure the intended outcomes of the programme should be included. Thus in the case of the Enterprise Zone evaluation (Public & Corporate Economic Consultants, PACEC, 1996), both directly related public expenditures on EZs such as tax revenue foregone from capital allowances, as well as public spending on infrastructure and land reclamation on the EZs undertaken under the auspices of different public spending programmes, were included. One difficult problem here is that of ascertaining how much of the expenditure under these other programme heads would have been incurred anyway even in the absence of the EZ initiative. Problems of identifying and measuring costs when programme bending is a significant aspect of a policy also arose with the evaluation of the Inner Cities Initiative (PACEC, 1991).

One other contentious area relates to revenue derived from sales of assets associated with a regeneration strategy such as reclaimed land or buildings owned by the public sector. Increased development values have also been used as a measure of the economic benefit from urban regeneration. Arguably such revenues should be regarded as a negative cost to be deducted from public expenditure; however, Treasury guidelines preclude this and suggest that revenue from asset sales be treated as a windfall gain rather than a deduction from public expenditure. The main arguments for this treatment of enhanced asset values are that: there is a risk of double-counting if the benefits are included elsewhere, say in time savings from improved transport facilities; displacement may occur if land value gains in one area are offset by losses in another area; and enhanced asset values may be embodied in the asset price in anticipation of the impact of policy.

Comparing Benefits and Costs

The above discussion has pointed to the numerous problems that must be addressed in establishing the outputs and benefits from urban regeneration strategies and the costs associated with such strategies. It is now appropriate to consider the options for comparing the costs and benefits of different programmes and initiatives. A major distinction to be made is that between cost-benefit indicators of performance and cost-effective indicators, as both find use in evaluation research. The former aims to express the outcomes of a programme or project in monetary terms suitably discounted over the project or programme lifetime. By contrast, in cost-effectiveness analysis outputs of a programme or project are expressed in terms of the expenditure necessary to generate a particular ultimate or intermediate output, e.g. cost per net additional job created, cost per unit of floor-space, cost per hectare of land reclaimed. In the 1990s cost-effectiveness analysis has been widely used to evaluate and compare different urban policies and regeneration programmes. In this type of analysis a particular focus has been on the cost per job performance measure. This methodology is not designed to compare resource costs with economic benefits but rather to compare different geographically targeted programmes and initiatives with respect to their capacity to divert jobs from one area to another. However, even where the cost-effectiveness methodology is deployed there are difficulties when initiatives are concerned with improving the access to jobs for specific target groups, such as ethnic minorities or long-term unemployed, rather than merely creating new jobs in a particular geographical area. In these circumstances the evaluation must aim to assess the specific benefits to particular target groups within an area.

Given that most regeneration schemes aim to secure a complex 'basket' of economic, social and physical objectives where assigning monetary values to outputs is often difficult if not impossible, the conventional cost-benefit approach is generally of limited applicability. One approach is that of developing a cost-effectiveness 'balance sheet' incorporating both quantitative and qualitative output or benefit measures for different components of the regeneration strategy against the costs incurred which may be at current or constant discounted prices. Table 10.2 illustrates this approach with examples of outputs and impacts for the different programme components.

Table 10.2 **Cost-effectiveness balance sheet for evaluating urban strategy**

Net additional costs	Net additional benefits
1. **New enterprise formation**	**Outputs** – Starts/survival rates
Public expenditure on the programme	**Impacts** – Jobs
Expenditure by other partners	GDP
2. **Inward investment**	**Outputs** – Investment by firms/new projects
Public expenditure	**Impacts** – Jobs
Expenditure by other partners	GDP
3. **Business advice**	**Outputs** – Percentage of firms failing
Public expenditure	**Impacts** – Jobs
Expenditure by other partners	GDP
4. **Improving skills and knowledge**	**Outputs** – Qualifications secured
Public expenditure	**Impacts** – Jobs
Expenditure by other partners	GDP
5. **Improving access to opportunity**	**Outputs** – Positive outcomes on target groups
Public expenditure/expenditure by other partners	
6. **Quality of life and social improvements**	**Outputs** – Number of users, housing improvement, education/health gains, crime reduction
Public expenditure/expenditure by other partners	
7. **Improving the physical environment**	**Outputs** – land taken up for development, floor-space created/refurbished, environmental improvement
Public expenditure	**Impacts** – Jobs
Expenditure by other partners	GDP
Total spending	**Total outputs/impacts**

Source: PACEC (1999)

An alternative way of presenting the information in Table 10.2 is to show the combination of outputs or benefits that can be secured for a given amount of total spending (public plus other). Table 10.3 gives an example from the evaluation of Inner City Task Forces.

Table 10.3	**Package of benefits/outputs per £10,000 expenditure**

Unit of cost	Outputs/benefits
Every £10,000 of net expenditure	1 job (2 job years)
	2 training places (1.3 training place years)
	40.6 small firms assisted
	12 residents receiving community benefits
	0.75 sites improved environmentally
	£3.8k spend on improving the physical environment

Source: PACEC (1991)

In addition to the outputs indicated in Tables 10.2 and 10.3 there will be other achievements flowing from an effective urban regeneration strategy which might usefully be documented including, importantly, the development of appropriate institutional structures within policy can be delivered.

An Overview

The detailed discussions in this chapter have introduced a number of concepts, procedures and methods for the measurement, monitoring and evaluation of urban regeneration. Irrespective of the purpose of urban regeneration, and despite the many changes that have occurred in terms of the detailed form and structure of policy, one thing is certain, and that is that the participants in urban regeneration will always be required to demonstrate that they are aware of what they intend to do, how they intend to set and achieve their objectives, and how they will measure, monitor and evaluate the actions which they have taken.

It should also now be clear that the basic rules, regulations and procedures used in relation to measurement, monitoring and evaluation do not change significantly over time. The Treasury, the European Commission and private funding institutions may use various terminology, but the principles remain the same. The key issues and actions in connection with measurement, monitoring and evaluation are summarised below. However, it should be noted that all evaluation exercises need to be customised to reflect the nature and scope of regeneration programmes and initiatives and the opportunities and circumstances in each location.

Key Issues and Actions

- Develop an understanding of the requirements of funding bodies and learn the terminology.
- In developing a strategy incorporate a framework of measurement, monitoring and evaluation.
- Make sure that you require all participants to maintain records in accordance with specified requirements.
- Define the milestones and set strict dates for interim reports.
- Develop appropriate procedures for measurement, monitoring and evaluation, and make sure that participants understand these procedures.
- Collect all the direct survey information that is required at regular intervals.
- Continue to collect all indirect information from external sources that help to demonstrate the progress of a programme or project.
- Do not leave evaluation to the end of a programme or project, start this process at an early stage.
- Use the information collected to review and revise a programme or project.

Glossary

The terms used in monitoring and evaluation are listed below:

additionality – the extent of change resulting directly from an action compared with what would have occurred without the action (in European Union programmes this term also has a very specific legal meaning in terms of the regulations governing the Structural Funds).

appraisal – the process of defining objectives, examining options weighing up the costs and benefits, and anticipating the likely outcomes of a project or programme (sometimes called ex ante evaluation).

assessment – a general term for a process of evaluation or estimation.

baseline – the starting-point for a project or programme against which change is measured.

benefits – the outputs of an activity, these can be direct or side-effects.

costs – the inputs to an activity, normally expressed in financial terms, but can include other inputs (for example, contributions in kind).

counter-factual – a statement of what would have taken place anyway without policy intervention.

dead-weight – outputs that would have occurred anyway, without any policy intervention or action.

deliverables – the outputs which a plan or project or programme intends to achieve.

delivery plan – a plan setting out what a project or programme intends to achieve, when, where and at what cost (see also *milestones*).

displacement – the extent to which an additional desirable output reduces or prevents outputs elsewhere.

effectiveness – the extent to which the objectives or goals of a policy, project or programme are achieved.

efficiency – the ratio of an output to the resources used to produce the output.

efficiency testing – the examination of alternative ways of achieving the same effect with fewer resources.

evaluation – the process of checking (after implementation) to see how far objectives have been achieved, what resources have been used and what outputs have been produced; it is also helpful to identify good and poor practice and to isolate what lessons can be learnt for the future (also called ex post evaluation or ex post review).

ex ante – an initial view (see appraisal)

ex post – a retrospective view (see evaluation).

externality – a benefit or cost that is not generally (or fully) reflected in market prices such as pollution or congestion.

gearing – the matching of inputs from one source by inputs from other sources.

grossing up – estimating the full impact of a programme by weighting a sample of initiatives by all initiatives.

impact – the net additional effects of a project or programme.

induced effect – the indirect benefits arising from income or expenditure in an area.

intermediate output – an early output produced as a result of an intervention, the value of which is not always quantified therefore there is usually a need to measure final output (see *outputs*).

leverage – the additional non-public sector input that is induced as the direct result of policy intervention; normally applied to private sector activity.

market failure – anything that prevents markets from operating freely on the demand or supply sides, such as institutional restraints or restricted competition.

milestones – key events, targets and other indicators, often dates are attached to milestones.

multiplier and linkage effects – the second round effects or the level of economic activity resulting from an initial injection of expenditure.

objective – a statement of what it is planned to achieve and by when, for example, intermediate objectives are staging points in the achievement of final objectives.

outcomes – the wider effects or impacts on an area or sector of a project or programme.

outputs – the early results of a project or programme (sometimes called final outputs).

outturn – the actual as against the estimated level of input to a project or programme.

participation or take-up – the number of individuals or firms that use an initiative or service.

policy – an objective (or set of objectives) together with a general specification of how it is to be achieved.

programme – a group of interventions often linked together over time and encompassing more than one project.

project – a single intervention or discrete, one off, form of activity.

side-effect – effects, beneficial or otherwise, which do not contribute to the final objective of a policy, programme or project.

substitution – this occurs when subsidised resource inputs take the place of unsubsidised resource outputs (see also *displacements*).

sustainability – the extent to which an achievement can be maintained, this term is sometimes used in place of the term environmental sustainability; this usage has a different interpretation from that employed in evaluation terminology.

synergy – the process by which programmes, partners or activities interact, often achieving more than the individual elements would.

target – a quantified objective.

weighting – a technique used to combine a number of outputs into a single measure of output even though they cannot be valued in money terms.

References

Department of the Environment (1995) *Single Regeneration Budget Bidding Guidance*, Department of the Environment, London.

Department of the Environment (1989, 1996) *Interim and Final Enterprise Zone Evaluations*, PACEC, Cambridge.

Department of Trade and Industry (1991) *Evaluation of the Task Force Initiative*, PACEC, Cambridge.

Department of Trade and Industry (1992) *Inner Cities Initiative Fact Sheet*, Department of Trade and Industry, London.

HM Treasury (1988) *Policy Evaluation: A Guide for Managers*, HMSO, London.

Trafford Park Development Corporation (1995) *Strategy*, Trafford Park Development Corporation, Manchester.

11 Organisation and Management

Dalia Lichfield

Introduction

The purpose of this chapter is to identify the ways in which urban regeneration may come about and to demonstrate how good organisation and management may increase the likelihood of a successful outcome. It is therefore written with an eye on the roles of the project initiator or manager and the project team. The chapter has particular regard to the Single Regeneration Budget, whose *Bidding Guidance* (Department of the Environment, 1995) had a profound effect on the approach to urban regeneration, but it is not simply an exponent of SRB and goes beyond it in several respects. Despite the fact that many of the examples are drawn from the experience of SRB, the approach adopted in this chapter should remain relevant irrespective of changes in policy.

Many of the issues discussed in this chapter overlap with the more specific material developed in the thematic chapters. But whilst previous chapters have been concerned with a variety of individual topics and themes, this chapter is structured essentially around the process of urban change, where these various themes interact. The understanding of that process provides the basis for a step-by-step consideration of the management of urban regeneration. The chapter has three underlying themes:

- the need for the many participants in urban regeneration to share knowledge (facts and concepts) as well as action;
- the importance of identifying who experiences existing conditions and who will benefit from the proposed changes;
- the importance of an integrated strategy and of resources as a key ingredient.

The methods and procedures advocated in this chapter are designed to increase the awareness of these themes among project management and partnership teams. Attention is given to the common dilemmas faced by initiators and managers, the relevant forces to be reckoned with, and the extent to which obstructions may be removed and positive action encouraged.

A good idea for urban regeneration will not implement itself. The pathway from good intentions to concrete action is often paved with obstacles and difficulties, many of which can best be overcome through a combination of determination to progress a scheme coupled with clear and decisive management. It is therefore essential to ensure that management and associated issues are:

- considered as a central concern;
- built into the design of a programme or project from the outset.

The primary goal of management is to create an organisation that will enable the sharing of knowledge between participants and facilitate agreement on strategic vision. The management structure should reflect the approach to planning pre- and post-approval and to implementation of regeneration.

The planning process (and the organisation) is envisaged in three cycles:

(1) A preliminary or provisional scoping of problems, potential objectives, and 'actors' on the urban scene – this allows the project initiators to set up the core organisation and also to identify additional partners and key issues that require more detailed study or communication efforts.
(2) A strategic planning phase in which all the 'stakeholders' are brought together, the various impressions and assumptions made in the first phase are verified, and a specific strategy is forged in agreement between the partners, for submission to the authorities who control the budget (Government Offices for the Regions in the SRB regime);
(3) A detailed planning phase following approval by the authorities.

The following sections of this chapter deal with eight major topics.

- Phase 1: Scoping
  - initiating urban regeneration;
  - developing an initial view of the urban regeneration programme.
- Phase 2: Finalising the organisation and preparing the strategy
  - identifying actors and constructing a partnership;
  - preparing an urban regeneration strategy;
  - preparing the formal bid or proposal.
- Phase 3: Into action
  - office administration;
  - accountability and responsibility;
  - preparing detailed project plans.

Phase 1: Scoping

Initiating Urban Regeneration

Two issues are of particular significance in the pre-approval stages of urban regeneration:

- identifying the basis for an urban regeneration programme or project and gaining local support for the proposal;
- appointing a project manager and setting up the partnership.

The first of these issues has been discussed in a number of earlier chapters (especially Chapters 1 and 2) and only certain aspects of it are raised here, linking into the discussion of the second issue in greater depth.

Identifying the Basis for Regeneration

As was noted in Chapter 1, the factors that stimulate the development and application of urban regeneration programmes and projects vary from place to place. However, in general, the primary reason for urban re-generation is to respond to urban changes that have detrimental effects – either traumatic such as an industrial closure, or gradual such as the econ-omic and physical decay of an area. The urban scene (economic, physical and the human scene) is constantly evolving and adapting through market forces within the existing framework of public policies. Additional public intervention, dedicated to urban regeneration, is called for where market forces do not effect regeneration or, while regenerating some aspects, leave certain groups in the community poorly off.

It is therefore not sufficient to note the indicators of deprivation at a single point in time but to view the process of urban change and assess its likely outcomes, if additional public intervention were not provided.

Once the process of decline/regeneration in a given area is understood, the regeneration team should also be able to identify the forces at work and which of them could be influenced or countered by the regeneration strat-egy. At the same time it is possible to select those that should be party to the project.

Against this background the nature of the project, and its organisation, can be formed.

On to a Sound Start: Appointing A Project Manager

The initiators of an urban regeneration programme or scheme (local auth-ority, TEC or other organisation) face their first dilemma with the appoint-ment of a 'project manager'. Most initiators realise that, ideally, the manager should have:

- the intellectual and professional ability to grasp the complex issues of decline and regeneration in a holistic or integrative way;
- the creative ability to innovate and break traditional moulds;
- the interpersonal skills to inspire and to facilitate collaboration among different individuals and organisations;
- the track record, rank and status to command resources and action;
- professional skills related to the main thrust of the programme (econ-omic, housing, social, etc.)

The role of this person is to build up the organisation and manage the urban regeneration programme or scheme, creating consensus and a useful vision, and taking the activity forward to the point of funding and implementation. Developing an urban regeneration programme is a major project in itself and in an ideal situation requires both the support of a champion (or champions) and the skills of an experienced manager. It is essential at the outset to appoint both a competent chair and chief executive. These key people must be able to work together and with other partners.

However, in many cases the initiators cannot afford to allocate a person of such quality to the task of project management before funding has been secured. This dilemma is common in the SRB funding regime, where no assistance is given for the pre-bid phase, while at the same time the bidding proposal is expected to provide sufficiently solid information to allow for the judgement of the 'value for money' of the proposal. In practice, therefore, initiators commonly choose one of two solutions:

- a highly capable director of an existing unit or department (usually in local government) takes on the regeneration project as an additional task, and it is then liable to receive only limited attention, being of a secondary priority;
- alternatively, a lower-level person is appointed to deal with the project on a full-time basis, but their capabilities and status result in an unsatisfactory product.

In such circumstances the search for a suitable manager starts later, after funds have been allocated, but by then the initiator is bound by the package which secured the funds and potentially superior solutions may well have been lost.

There is no easy answer to this dilemma and circumstances clearly vary. Some changes in practice may also be worth considering:

- Central government might find that taking up a relatively small part of the capital programme budget for pre-bid funding is worth while (ensuring that it is spent on project management and planning rather than on packaging).
- A local authority may consider it worth while appointing a suitable manager regardless of central government funding, bearing in mind that the development of the project may bring benefits even if a specific bid for funding is not successful. For example, if local government heads share 'new knowledge and approach' this may influence their action in their capacity as city-wide officers. This in turn can affect the area of decline (for example, by transportation, affordable housing, location of business).

Once the decision to recruit has been taken, urban regeneration managers may be recruited from various sectors – both public and private – so long as they possess the five qualities mentioned above. The first four (integrative grasp, creativity, interpersonal skills, status) are to a large degree a matter of personality. The fifth may require 'horses for courses', subject to qualifications. But a word of caution: persons with highly specialised professional skills may also be narrow in outlook and, if so, would be more suitable as team members than as project managers.

An Initial View of the Urban Regeneration Strategy

Formal Requirements

Most regeneration programmes will be initiated with an eye on particular sources of finance. The finance providers will have their own requirements regarding the focus, and scope of regeneration, as well as accountability procedures. Specific requirements may be specified by, for example, by the SRB, English Partnerships, the European Partnerships, a local authority, private developer or charitable organisation. Central government, increasingly through its regional offices, plays a major part in setting the rules.

Although the precise definition and focus of regeneration policy has been subject to change in the past (Lichfield, 1984; 1992), a number of basic principles have evolved from City Challenge, SRB and, more recently, the policies of the present Labour government, that underpin the management of an urban regeneration programme or scheme:

- An integrated approach that brings together the many aspects of urban life has replaced the earlier single focus (property-led; training-led, etc.) approach. This integrated approach has been promoted by SRB guidance and later by the present Labour government's general planning policy guidance.
- Partnerships should be set up between several of the public and private sector bodies involved, of which the local authority must be one and preferably the leading force, since it is normally the single largest manager and spender in any area through its various departments.
- Schemes should demonstrate that they provide value for money.
- Objectives should be clearly stated and progress and outputs clearly monitored.
- Proposals should conform with an existing strategy for a wider area or should help to shape a more local strategy.
- Whilst SRB guidance indicates the general structure of partnerships and a checklist of objectives and outputs, it indicates that regeneration areas should exercise discretion and determine what suits its particular circumstances.

The Trigger to Success

These requirements are essentially sound and deserve to be maintained even if other aspects of SRB are abandoned. However, their effectiveness depends on the proficiency of the detailed application. If properly handled, this approach to regeneration could improve the way that ongoing local planning and development in the area and the city at large is carried out. It could break down the barriers between land-use planning, the planning of education and social services, economic planning and environmental management. It could become a more advanced and democratic version of corporate planning, reflecting open government and consultation. In addition, it could enhance the general performance of local government as the managers of the local scene, working together with the multitude of local actors and stakeholders, many of whom will be in the private sector.

However, in order for these activities to proceed and these benefits occur, all those involved must speak the same language, share an understanding of facts and processes, and possess the resources and delivery mechanisms. This is the trigger to success.

This trigger is, in many places, still lacking. The challenges for the project manager in this situation are:

- to grasp the scope of the issues in the widest possible way;
- to identify the variety of actors involved in promoting change and those that will experience its results, and to appreciate the extent of common ground and of any differences;
- to set up an organisation, process and management style which maximises the necessary sharing of knowledge and action.

The challenge is best met through an iterative process: a provisional 'scoping' is prepared by the manager on the basis of early briefings, before proposing the outline of the project and the selection of the key actors for the partnership of planning and implementation. The provisional findings are later extended or amended in collaboration between the actors and by further study. In this way there will not be the need to amass information that may be of no use. Rather, in trying to clarify specific questions raised during the scoping, the project manager will appreciate what information is truly necessary for credible answers.

The main elements of the provisional 'scoping' and the preparatory investigation are:

- stating and defining the problem;
- understanding the process of change and identifying actors;
- developing an approach to regeneration;
- producing an initial proposal.

Stating and Defining the Problem

There are various definitions of decline and several systems of measurement of local deprivation. The majority of them, including the Department of the Environment Index of Local Conditions, are global indices based on general statistical data. These are useful in providing general guidance on the state of one area as compared to another, but are less helpful when attempting to identify the specific conditions and variations within a given area. Indeed, 'need' and a high index score are not prerequisites for SRB applications. The problem of decline is often in the eye of the beholder, and people's initial perception of the problem is rarely identical. Broadly speaking, perception is influenced, first, by personal experience of the situation (for an outsider it may be physical neglect, for an elderly resident there may be a safety problem, for another an unemployment problem) and, second, the perception of problems is affected by knowledge and interpretation (appreciating the range of problems experienced by others, understanding interrelations as, for example, between juvenile crime, policing and social and economic conditions).

Therefore, although individual impressions or those of an interested few may provide a starting-point, it is essential to be aware of the limitations of such perceptions and consciously seek to expand (in discussion with other observers and interest groups) the understanding of decline in that particular area: what is the process leading to it, what are the ensuing problems and – most important – who are the people that experience these problems, whether within the area or outside it.

Some people may define 'problems' as the causes of the current difficulties (for example, the decline of an industry, causing unemployment), but ultimately the problem is an impact as experienced by people (that is, unemployment or absence of services). Table 11.1 provides an example of different individuals and organisations affected by industrial decline in a given area.

Table 11.1   **Groups of people affected by problems**

Group	Unemployment	Personal safety	Congested high street
Local black males	105 unemployed		
Local elderly		15 mugged, 350 fearful	
Shop keepers			Reduction in trade and poor image
Local shoppers (including other groups)	Low and falling disposable income		Pollution, danger, rundown shops
Town-wide car and bus users		Many attacks on bus users at night	10,000 people daily

Understanding the Process of Change and the Participating 'Actors'
Understanding the process of change that leads to decline (as opposed to a 'snapshot' of the existing situation) is crucial for a successful attempt to reverse that process and to create a realistic strategy. When exploring the process it is necessary to focus on:

- the place of the area in the city's economic and social, as well as its physical, fabric and whether it plays a positive role, for example, in accommodating low-income people and marginal businesses (Lichfield, 1994a);
- why the area attracts its current activities and what chain reactions contribute to the current state of affairs;
- the extent to which the area demonstrates a stable state of poverty or progressive decline and deprivation;
- how much of that process is due to change through local actions and how much is the result of national or other forces outside the control of the local partnership?

Such insights will prove essential when considering the desired regeneration strategy.

Understanding the process of change will bring to light the full range of 'actors' or 'stakeholders'. It includes those involved in creating the change (from both public and private sectors) and others affected by the outcomes (individuals and organisations who experience either problems or benefits).

Developing an Approach to Regeneration

On the face of it, it is possible to suggest that regeneration simply constitutes the reversal of the problems as perceived by the relevant groups in Table 11.1, but there is more to it. The SRB *Bidding Guidance* suggests that regeneration can be seen as a shorthand term for sustainable regeneration, economic development and for industrial competitiveness.

It is generally accepted that regeneration ought to be sustainable rather than temporary, that is, an improvement that will lay the foundations for the long-term life of an area rather than a temporary improvement related to injection of money or a public relations hype. Sustainable regeneration, if it does not take place spontaneously, requires intervention to induce a change in processes, especially a reversal of the process of decline. Such intervention can take various forms (including physical means of change, such as a road that changes the accessibility of an area and thus its economic prospects, or economic initiatives which start investment interest rolling, or social means such as education).

The function of government in relation to the private market is a familiar dilemma when suggesting interventions. The ongoing policies and activities of central and local government can be regarded as interventions affecting the spontaneous market processes. Several are already likely to be in place and to affect an individual area. The project manager will have to be familiar with all of these interventions, since the urban regeneration initiative will be most effective if it can complement them and, more ambitiously, co-ordinate them.

What processes can one control? Understanding the forces of action helps to differentiate between external forces that are outside the control of the organisation, intrinsic factors in the area which may not be given to a change (for example, location, where it is difficult to change the relative accessibility of an area) and controllable factors which can be managed by the organisation.

The above process of assessment and project development will enable the project manager to confront a number of crucial questions:

- What is really aimed at – is overall human suffering the main concern?
- Who is the activity intended to assist – which groups are inside the target area and which outside it, and what weight is put on the depth of impact or on the numbers involved?
- Is it possible to identify the current and potential roles of the area in the social and economic fabric of the town and will it maintain its positive functions (for example, supplying low-cost accommodation for low-income people and marginal businesses that do have a place in society)?
- Is the area simply poor or is it descending down the spiral of decline?
- Has the analysis, and the actions proposed, brought together all those who experience the problems of the area and those who can help to bring about change?
- How can a baseline of the real problems be established, against which realistic improvements can be anticipated and then monitored and measured?

Recording Provisional Proposals

It is likely that several of those involved in an urban regeneration initiative, whether among local communities or official agencies, will already have developed proposals for improvement. These proposals will be based on varying degrees of local knowledge, which it is important to capture, but may or not provide the best solutions. Proposals may comprise an overall strategy or may simply be a collection of fragmented projects which will have to be tested later against an integrated regeneration strategy. When recording and assessing these proposals it is useful to record their origins, any particular interest groups that will benefit from them and whether any others will be disadvantaged.

Since the first challenge for a project manager is to promote common ground between participants, it is essential that differences of view are clearly recorded. It is also useful to trace their origins, as differences of views on 'what is the problem' or on 'the best solution' may be due to difference in knowledge of facts or processes, or due to self-interest. Proposals may enjoy general or only partial support from other groups or agencies. Wide support is a good omen, though not necessarily a proof of the soundness of a proposal!

The preliminary review in Phase 1, conducted by the project manager on the basis of readily available information, is a valuable starting-point but is likely to demonstrate that a proper understanding of the issues and agreement on possible solutions will require a much greater and more methodical effort in Phase 2, in particular:

- bringing together the relevant actors (stakeholders);
- collecting essential information to verify impressions and assumptions;
- creating an agreed and credible strategy.

Phase 2: Forging the Organisation and Preparing the Strategy

This is the strategic planning phase in which all the 'stakeholders' are brought together, the various impressions and assumptions made in the first phase are verified, and a specific strategy is forged in agreement between the partners, for submission to the authorities who control the budget (Government Offices for the Regions in the SRB regime).

Managing the Actors and the Partnership

The Range of Organisations and Actors

An essential task for the project manager is to ensure that all appropriate actors are involved in the process. In Phase 1 the manager will have gained an impression of all actors and stakeholders; and this can assist in the selection of those that should participate in the planning and implementation efforts.

The list of actors or participants in an urban regeneration initiative should include all those whose regular activities and policies affect the fortunes of an area (including the local authority, voluntary organisations, local businesses and local consumers); those who provide resources or budgets for a special initiative (for example, central government agencies and major investors) and all the groups (both within and outwith the area) that will experience the impacts and therefore support or object to the initiative. This section of the chapter expands on the relevant community groups, whilst other chapters discuss economic and financial actors (Chapter 4) and employment and training (Chapter 7) organisations.

Establishing a list of actors may be assisted by the use of the matrix which is presented at Figure 11.1. This diagram identifies actors who are directly involved in interventions (such as developers, service providers, major employers and also government) and actors who experience the change and whose personal behaviour may then reinforce the change (for example, local youths who will have new challenges and so commit fewer offences, shoppers who respond to improved environment and make more use of local facilities); and (on the vertical axis) actors who are active inside the area (such as local employers involved in a direct intervention or local youth experiencing its results) and actors from elsewhere (including government changing a general policy or employees outside who lose their jobs because their employer has transferred the business from its old location into the regeneration area). This latter set of actors are normally not considered to be involved in the process of regeneration, but none the less experience its external effects, or externalities.

Figure 11.1   **Identifying and Selecting Actors**

ACTORS	INTERVENING	EXPERIENCING EFFECTS
WITHIN AREA		
OUTWITH AREA		

In considering the list of actors it will become evident that some experience both effects (for example, retailers experiencing the results of traffic interventions) and also intervene (when in consequence they close down or invest in new shops). This form of analysis helps to expose the chain-effects that form part of the process of regeneration.

In selecting the actors who can affect change, it is wise to differentiate between changes that are within the initiator's powers to implement, changes which other forces and groups could implement if collaboration was established, and changes which are outside the control of either (for example, national and international economic policies). Selecting actors often raises the issue of accountability in urban regeneration, and two questions must be answered:

- Is the person nominated by a body that is truly representative of the full range of interests (particularly in relation to local groups with unelected representatives or with a majority representation only)?
- Does the representative have sufficient authority to commit his or her organisation to a specific cause of action (this is particularly true of official bodies)?

It is likely that the number of people who should participate will be larger than may be convenient for groupwork on planning and management. The organisational structure should cater for this, allocating people different roles while also ensuring the provision of platforms for dialogue and for a general exchange of views.

As the programme boundaries and its strategy are re-examined during the next iteration of the planning process, it may be necessary to re-examine and adjust the list of actors.

Preparing the Strategy

Having conducted an initial review of the problems and opportunities evident in an area requiring regeneration, and having developed a provisional programme and partnership, the next step in the management of an urban regeneration initiative is to prepare a detailed strategy. There are five main stages in this process:

- developing an integrated approach to the definition of the key objectives and the strategy;
- generating agreement on integration;
- agreeing the principles of a strategy;
- identifying potential resources and other requirements;
- defining the boundaries of the initiative.

Whilst in an ideal world the preparatory work as outlined in this chapter provides a logical and economical pathway to the management of the development of a strategy and the bid, in reality the management and planning process is sometimes disjointed or disrupted, and the stages do not follow each other in a logical sequence. Sometimes it is necessary, for example, to identify the actors and form the partnership before attempting to define the problem. Nevertheless, if a cyclical approach is adopted (a provisional Phase 1 and an actual second round as Phase 2), then there is scope to modify provisional structures and ideas as the knowledge increases.

The importance of sound preparatory work has been emphasised in many official documents such as SRB *Bidding Guidance*. Under the SRB bidding regime many proposals did not go beyond the preparatory level of work and the initial views and ideas in fact became the 'bid package'. The reason was that the initiators of regeneration often felt unable to invest substantial resources in proper planning in advance of their bids being accepted. Thus while SRB *Bidding Guidance* calls for value for money, many proposals fail to fully explore or compare alternative strategies and, therefore, have not necessarily adopted the strategy which provides the higher value for money. Once the bid has been approved, detailed planning tends to proceed on the basis of the package which won the bid. Alternative strategies thus do not see the light of day. This anomaly might be avoided if funds were provided for pre-bidding planning or if competitive bidding were abandoned.

But credible planning can also be facilitated where local authorities are prepared to find the resources themselves or in partnership. The work programme advocated in this chapter assumes that this is the case. The resources required can be quite modest: there is no need to farm out the entire exercise as an expensive consultancy project, since an experienced consultant can act in a 'hand-holding' capacity to the local authority and the partnership staff. Some would do so at a reduced fee scale, provided that the partnership undertakes to employ them if and when funding is available for detailed planning.

An Integrated Approach

Having put together a team, the project manager will have the ongoing challenge of developing the common ground between them. They should therefore share in a discussion or debate, as was introduced in Phase 1 above, about the nature of decline and the meaning of regeneration in the area. Before embarking on the preparation of the regeneration strategy proper it would also be wise for the team to explore the boundaries of their area and programme as well as the concepts of 'strategic planning' and of 'integrated planning', since a degree of shared approach to these issues is essential for their collaboration. This discussion of the basic principles of urban regeneration and methods of approach should also be extended to other key actors, in the first instance, and the wider community (residential, business and institutional) in an area at a later stage.

The importance of an integrated approach to defining objectives and the strategy can be seen to be evident from the earlier sections of this chapter (Lichfield, 1994a) and reflects two issues:

- the complexity of the phenomena one is dealing with requires of itself an integrative or holistic outlook on the world in order that processes of decline or regeneration can be properly understood;
- regeneration involves a much wider variety of organisations as well as of informal groups of people and other powers, all with their different cultures and objectives; bringing them together under the umbrella of regeneration requires special skills and efforts.

Generating Agreement on Integration
The relevance of this debate to project management is clear: integrated analysis and understanding of the urban scene and an integrated evaluation of the options will bring to the surface diverse interests. A well-managed discourse will enable each of them to appreciate the presence and concerns of the others. There are also negotiating techniques that enhance understanding, such as 'role playing' and the Lichfield Budget Line (Lichfield, 1994a).

The practical expression of an integrated approach revolves around eight types of integration:

- geographic integration – addressing an area as part of a larger city and region, since both the problems of the area and the desired improvements result, to some extent, from conditions in other parts of the city;
- integrated data – collating data from different sources, including the education, health and social services departments and the police to create a unified information system which will provide a comprehensive picture available to all concerned;
- integrated interpretation of the scene – understanding the interactions between education, income, housing and other factors and the importance of chains, for example, the need for training and information about jobs and child care facilities and adequate transport and supportive family (to bring a young mother into the workforce);
- integrated planning team – teamwork, involving the various actors and professional disciplines;

- integrated plan-making and implementation – incorporating consider-ations of feasibility and delivery mechanisms, using plan-making itself as an agent of change in attitude and action of the actors on the scene;
- integrated funding regimes – using funding from different sources, for example, SRB and Objective 2, in a complementary way;
- integrated policies – consistency between the policies of different gov-ernment departments and down the hierarchy from central to local government and area administration;
- integrated action – collaboration between all those involved in implementation.

City Challenge and SRB opened the door to an integrated approach, but doubts remain as to the full adoption of this approach and to the avail-ability of expertise and support for its implementation in the field. The desirable mode of operation, however, transcends particular government programmes.

The Principle of a Strategy

Strategy is an overall framework for action which indicates, in general terms, how one's aims might be advanced. A good strategy is one in which creative ideas are allowed to develop within that framework on the basis of:

- an understanding the powers and processes which affect the area;
- an ability to co-ordinate the available resources and intervene in these processes.

A strategy consists of a package of five key elements:

- aims and objectives, clearly stated;
- understanding of processes and of 'our' and 'the enemy's' strengths and weaknesses;
- appraisal of resources (not only financial) available;
- creative ideas which go beyond the trodden path;
- selection of a realistic course of action in outline terms, allowing 'field actors' local discretion and providing value for money.

A combination of all five elements provides a strategy, as opposed to a set of long-term objectives. But a strategy should not be cast in stone and may evolve with experience. Indeed, part of a strategy is constant reassess-ment and adaptation to the evolving scene (see Box 11.1 on strategic decisions). Clearly a well-informed decision will better serve the all-round public interest in regeneration.

Value for money means that for a given amount of money the selected strategy provides higher-value outcomes than the next option; or that it provides the same outcomes for lesser cost. Value is not only measured in financial terms, but also in social, economic and environmental terms. This principle, promoted by SRB, is strongly endorsed by the 1999 UK govern-ment. Value for money in strategic planning therefore requires the pre-paration and comparison of options.

Box 11.1   Strategic decisions

A strategic decision is not simply about large quantities, but has the following attributes:

● It can result in a changed strategy – changed objectives, strengths and resources or the adoption of a course of action.
● It is a complex decision.
● It is not a routine decision.

For example, visualise a strategy aiming at full employment, relying on the private market. After a while we observe that the market does not have the resources to deliver. Now we can change the component of resources in the strategy and inject public finance; but understanding the processes indicates that this would lead to inflation, an undesirable outcome. So we can try revising our strength instead and put more effort into education and training to increase our relative advantage over competing countries. But are we fully aware of the strengths of the enemy and the routes of competition? If we have miscalculated, we may need to revise the strategy again and perhaps modify our aims/objectives. Such decisions will only take place once in a while and be taken by the full board of the organisation.

Identifying Potential and Resources

Almost every urban area – from the richest to the poorest – has some potential for improvement in the quality of life and activity of its residents, or in the contribution it makes to the wider urban or regional public. Such potential exists in physical, economic and human terms. It may be expressed in underutilised property or neglected environment, in unrealised economic advantage, or in jobs, workers and training that never come together. On occasion it is present even as unused financial resources. The potential may dwell in local organisations and authorities that are inefficient or unco-ordinated and, above all, in people who, by being unemployed, underemployed or uncreative, do not use to the full their human potential.

The urban scene, with its underutilised resources, involves a very high turnover of financial and other resources year on year. Put together they amount to many times the value of dedicated urban regeneration budgets. Even a small degree of improvement in the effectiveness of the overall budget with regards to urban regeneration objectives would have the output equivalent of a significance comparable with the dedicated urban regeneration budget (see Box 11.2). A strategy that aims to improve the management of available resources requires a considerable degree of careful thinking and investment in the people whose plans, decisions and implementation are capable of producing that improved management.

Defining the Boundaries

The word 'boundary' is used here to mean both the 'remit' of activities and the physical boundaries. The boundaries of a regeneration project may

---

Box 11.2   The Ashkelon experience

In 1980 a major part of the town of Ashkelon, Israel, was at the bottom of the list of areas of decline, beset by general despair, lack of confidence and apathy. The urban regeneration strategy for the area was to restore confidence and initiative by assisting the local authority, local community organisations and people as individuals to improve all manner of underutilised resources, to manage them wisely, and to undertake preventative measures against decline so that more costly remedial measures will not be necessary in the future.

The strategic planning process, with wide public participation, performed an educational role for both officials and other actors in making methodical links between problems, processes, the evaluation of solutions and the management of resources. These officials developed new outlooks and techniques which they applied to their regular work. The place has been gradually transformed and managed to sustain and advance the good work despite the exit, after ten years, of the special regeneration fund. This project was selected by the Israeli Ministry of Interior as an example of good practice.

Source: Lichfield (1986)

---

have been clearly given from the outset, or may have remained controversial during the scoping exercise. Even if apparently 'given', it would be useful to consider the variety of boundaries which will in fact come into play in the process of regeneration.

Boundaries may be defined for the range of activities which the programme should encompass (for example, housing and training only, or also health and public transport). The implication of an integrated strategy for regeneration is that a fairly broad boundary should be adopted. This boundary should emerge from the initial scoping of problems and solutions.

Boundaries which define the physical area may not be uniform. A variety of physical boundaries is related to the following aspects of regeneration:

- Community boundaries could be for a community of neighbourhood or for community of interests. Most regeneration programmes relate to a local community which may exist to varying degrees on the basis of neighbourliness; at the same time these people partake also in communities of interest (that is people with whom they share interests, often over a wider geographical area).
- Operational boundaries will also vary. Most regeneration programmes are composite and involve several economic activities and services, each having different catchment areas or administrative boundaries. Thus any change in the existing pattern of activities, services or rules regulating them has to take account of the boundaries within which this change will take effect. A dilemma may arise when an improvement required for the regeneration target area would introduce better services within a wider boundary which was not designated to benefit from that budget. If this is not acceptable, the answer may be to change the service boundary.

These considerations are particularly relevant if one wishes to:

- develop a realistic view of the sense of community;
- address regeneration in the context of city-wide planning and management;
- create a shared data base for several of the actors involved in regeneration.

In most regeneration areas it is possible to distinguish between the target area proper and surrounding areas which contribute to the problems and possibly to the solutions. There may also be areas outside which will be affected by the area-based programme. Thus one may differentiate between the:

- 'target area' in which most benefits should be concentrated;
- 'study area' which includes the wider context, possibly city wide;
- 'action area' where intervention will affect the target area, and may take place in or outside the target area.

For example, a wider study may show that there is scope for employment or a superstore elsewhere with a good bus service, and residents of a deprived target area would benefit from it.

Preparing the Formal Bid or Proposal

A provisional view of issues and potential strategies should have been gained during Phase 1. Now is the time to share the knowledge and extend it, develop new ideas, confirm provisional views or modify them, and gain wider support for the project. There are eight general stages in this process, each of which needs to be carefully managed and to be the subject of agreement by all involved in the process. These stages are:

- collecting additional data and generating the evidence necessary to support the bid;
- confirming the aims and potential processes of regeneration;
- setting the context for the proposal;
- identifying the resource requirements;
- preparing the strategy options;
- evaluating the options and selecting the strategy;
- identifying the exit strategy;
- setting the basis for monitoring and final evaluation.

Data Collection and Analysis
General information and precise data which is available should be shared by all the actors so that their judgements are all equally informed. Among the considerations for the project team is the requirement for additional data. Is there sufficient knowledge, and is that which is available effectively assimilated by all involved in planning, deciding and acting?

In the preparatory work the project team will have collected available data on the issues mentioned, but the move towards generation of fresh data must be carefully considered since it normally involves higher expenditure. Data is only worth collecting if it clarifies specific questions which the team faces when interpreting the current situation or deciding their choices for the future. The tendency to collect and generate masses of data in advance can be very wasteful. On the other hand, those involved are not always aware that they should be seeking information and it is the project manager's role to raise consciousness of the main issues and the information needed to clarify them and to assess alternative solutions.

Ideally data gathered from various sources would be organised as a geographical information system which includes physical, social and economic data, and would be available for all to use. This is a big undertaking and may not be possible within the project. Forward-looking local authorities may, however, have city-wide data on a geographical information system (GIS) basis and the possibility of linking into such a system is worth examining.

In addition, the team should revisit the initial assessment of the nature of decline in the area, the problems identified and the people who experience them and it may require more information for that purpose.

Confirming the Aims and Potential Processes of Regeneration

The meaning of regeneration and its aims should be discussed and confirmed, both among the team and with the partners. Potential processes of change should be considered, bearing in mind what changes could be brought about by the activities of the organisation. This may lead to the adjustment of the matrix of actors and should result in greater mutual awareness among the team.

The aims of the initiative should be expressed in general terms, as distinct from objectives (which are more specific targets for a project or activity and will be set later with consideration of their feasibility). The range of existing project ideas could also be introduced to the team early on, although it would be premature to decide upon them before an analysis is made of costs and available resources.

Setting the Boundaries in Context

The team will consider whether it wishes to restrict itself to preset boundaries in terms of activities and in terms of geography, or whether it will tackle matters as and when necessary. In setting the boundaries the team should consider adopting different boundaries for the target area, the study area and the action area.

It is also vital to ensure that the potential links between the 'regeneration area' and city-wide conditions and policies have been highlighted. In practical terms, the future of the area may be affected by urban strategies devised independently by local authority departments of town planning, transportation, housing, economic development and education, as well as by strategies of major private sector bodies. Analysis of such strategies and liaison with them are at least a necessary precaution and potentially a way of reinforcing the regeneration strategy.

Resources for Regeneration

Resources include not simply financial resources but also other economic resources (for example, the potential for tourism) and human resources (including the capability of local people to invest their time in maintenance or in folk groups and the potential of local authority officers to function more effectively). Resources are normally of three kinds:

- external resources – these comprise both special funding from official or charitable bodies (for example, from SRB or the Peabody Trust) and commercial investments by the private market;
- redistributed resources – these could be sought where a change in funding priorities within local agencies (for example, the local authority) can bring greater benefits to the target communities, or where the transfer of an activity from one area to another can bring greater benefit (but always bringing into the equation those who will lose out in the other area);
- underutilised resources – in most areas there are some underutilised resources, for example: partly used or derelict sites and buildings; rivers or fields inaccessible to the public; a local university whose potential advantage in conjunction with local employers has not been exploited; local people with unproductive hours on their hands.

Among these three kinds of resource, the improved utilisation of local resources deserves special attention and could be an important plank in the strategy, for it is likely to sustain a long-term change rather better than a reliance on a single external injection of capital. For example, a local authority, which in most cities is the largest single business and controls considerable spending, may be able to continue its level of spending but improve its product. If such improvement has an equivalent product value of just 2 per cent, in an authority whose total annual budget is £200 million, it would be producing over a decade more than an SRB project budget.

Preparing the Strategy

The creation of the strategy amounts to a blend of:

- information and knowledge (processes leading to problems and opportunities; who is involved, resources and other factors);
- creative thinking and a broad range of examples (new themes and new ways of doing what has failed in the old ways; knowledge of good practice elsewhere);

- good judgement (in the selection of actors, of project boundaries and of the strategy) and sound evaluation (assessing proposals by way of all their impacts, not predominantly by financial returns);
- leadership and organisation (inspiring confidence and excitement; management which will deliver).

A creative idea or a clear overall theme, on a sound basis, should be easier to communicate and would gain more support than a package of bland or unrelated projects. This idea or theme is also more likely to provide a unifying framework for all to relate to. But one should not confuse the imaginative with the imaginary – an attractive vision which is not based on an understanding of the process of change, the range of issues and the availability of resources, could fail or be counter-productive.

Evaluating the Proposed Strategy

Ex ante evaluation assesses proposals before implementation, usually by comparison with alternative or 'do nothing' options. It aims to predict the outcomes of the proposed strategy options and assist in the selection of the most desirable strategy. The criteria for such assessment vary (see also Chapter 10) but ought always to reflect, in the final account, the impacts on people, including all those affected by the proposed changes (Lichfield, 1997). Ex ante evaluation is part of sound strategic planning, particularly when different interest groups favour different options and their diverse preferences should be tested against a comprehensive picture. A community impact evaluation (CIE) (Lichfield and Lichfield, 1992) takes account of all those groups in the community who are likely to experience beneficial – or adverse – changes as a result of the strategy. This includes the residential as well as the business community; the local as well as the city-wide community; present-day as well as future generations (Lichfield, 1994b) and it indicates who wins and who loses under each option, and makes it possible to cater for the entire community rather than for single issue and vocal interests. Such an evaluation provides a measure of the extent of public appreciation and support. The CIE technique also encourages transparency and accountability

Where ex ante evaluation is performed, the evaluation criteria should also be used for appraisal after the project has been implemented.

Setting an Exit Strategy

Urban regeneration agencies under the Urban Programme, SRB or the European Structural Funds are established to operate within a limited time and budget. Winding down at the end of the period could cause the loss of facilities and activities which people have become used to, resulting in an even greater sense of deprivation than before the project.

To minimise such damage an exit strategy should, ideally, be considered as part of the general strategy and its efficacy would be assessed by the ex ante evaluation of the proposals: will a proposed intervention lead to self-sustained improvement after the funding is withdrawn and how long does the proposed intervention have to be supported before it produces a self sustained improvement?

For example, the Ashkelon project (see Box 11.2) had as one of its main strategic objectives the generation of confidence and initiative. Although the project produced a sustained improvement after the project wound down, it did require some ten years to have that effect.

The exit strategy has to be considered more tangibly nearer the end of the operation. In addition to the overarching change, the exit strategy will be concerned with the continuation or cessation of discrete project activities. Essentially it is important to examine the past delivery mechanism, costs and outcomes of each project activity and then address the following questions:

- How successful has the activity been so far and is it worth continuing, modifying or ceasing?
- If wishing to continue, what resources and budget will be needed in future?
- Can the discontinued programme funding be replaced from other public sources, from charitable sources, or from self-financing activities including charging for a service?
- Is the current delivery mechanism of this activity capable of continued existence, with full responsibility and accountability, without the umbrella of the regeneration agency?
- If the current mechanism cannot continue, is there an established and accountable body which could take on the activity and, if it did, would the character of the activity be adversely affected, for example, would local authority control be too rigid, or would a business put profit before people?

Having addressed these issues a project manager can suggest the future programme of an urban regeneration initiative and the phasing-out arrangements.

Laying the Foundations for Monitoring and Evaluation of Outcomes
Monitoring of the project and its evaluation will be undertaken, during and following implementation, for several purposes including:

- informing the project team about the success or otherwise of their plans, so that the course of action can be corrected or the strategy revised; for this purpose monitoring and evaluation should be regarded as part of the planning process;
- accounting to the authorities and to the public about the operation.

However, the foundations for such evaluation have to be laid during the planning phases, for two reasons:

- so that the criteria used would be related to the criteria ex ante, when predicting the outcomes of the proposed strategies;

- to ensure the availability of baseline data against which to measure the degree of change.

The methods of such monitoring and evaluation are discussed in detail in Chapter 10 of this book.

Negotiations with the Funding Bodies

This chapter has already referred to funding bodies having specific requirements and the need to ascertain these from the outset. Once proposals are beginning to form it is wise to consult again with these bodies to build up a mutual understanding with them and avoid the risk of them rejecting the final proposals. The project management and planning team thus has two sets of clients: those authorising the development, and the communities involved in the area. The right time for negotiations will vary according to the bodies, procedures and the progress in strategic planning, but there is a need for contact before the strategy is finalised in Phase 2.

Phase 3: Moving into Action

This phase follows from acceptance of the strategy as outlined in Phase 2. The project office and procedures are set up and the strategy is translated into detailed projects for implementation.

Setting up Office

This section does not go into detail on these matters, but highlights three aspects which the project manager should consider.

The management and planning team should:

- identify necessary skills. In addition to the administrative staff, the following skills normally have a useful input: economics, social studies, police, engineering and transport; planning and urban design, development appraisal, legal. Particular circumstances may require housing managers, educationalists, youth workers, etc.;
- consider the use of in-house or external professionals, and where to find them;
- define the roles of the core team, additional professionals, and lay participants;
- assess the need for continuing professional development – what is needed, different modes of improving skills, and how to organise training.

Operating premises will be required and the following issues are important:

- the choice of an office location (on-site/off-site; access and parking; image, etc.). If the project is area based, consider the value of an on-site office to maintain good contact with local people and provide a first-hand link with the area. Space for the display of plans and ideas and for group meetings is valuable;
- the office functions and space requirements (for example, workstations, exhibition, meetings, public facilities; storage);
- equipment and standards (high or low standard and the image and cost implications).

Budgeting for the management office should be considered at the outset, including:

- capital costs (for establishing the office and purchasing equipment), overheads, and operating costs;
- book-keeping and accounting procedures.

Setting Accountability Procedures and Responsibilities

Accountability can be considered in two contexts: accountability to a recognised authority and accountability to the public. The essential components of a good and accountable management system include the definition of responsibilities; planning and budgeting; financial control and reporting; tendering and contracting; staff hiring and attendance. The regeneration agency is likely to be working to requirements set by its main financing body, such as the SRB, the European Community or a charity. Unfortunately, accountability requirements vary between organisations, and projects which have joint funding may find themselves duplicating effort. It is advisable to study the requirements in advance and negotiate a unified system with the funding agencies. Apart from official requirements, the regeneration agency may wish to establish its own high standards of operation and accountability such as Total Quality Management or other management practices.

These technical accountability systems apply equally to any public or private agency. However, the regeneration agency must look also to another client – the public in the target area. Public accountability takes on a different meaning here and would use different techniques.

The overall responsibility for the standard of operation of an urban regeneration scheme and for its accountability rests with the project manager, but experts will have to be hired from the outset to establish the finance and management accounting systems and to help prepare annual reports.

Monitoring the progress of an initiative for purposes of administrative accountability is discussed in Chapter 10. In addition to the technical and procedural matters involved in establishing a system of monitoring, it is also essential to ensure that the project team and all the participants involved in an initiative are provided with adequate training in the collection and analysis of monitoring information.

Preparing Detailed Project Plans

Urban regeneration projects are derived from the overall strategy and the method for their preparation repeats many of the essentials of regeneration planning, but with a sharper focus on:

- objectives and options;
- feasibility and implementation;
- impact assessment and comparison of options.

From Aims to Operational Objectives and Impacts

A project is normally associated with one of the key aims or objectives of the programme (employment, housing, etc.) and there are formal requirements to specify measurable objectives and milestones towards its achievement. However, similar results can sometimes be achieved via different routes, which make up the options to be considered. There may be substantial differences in costs, achievements and side effects between options and it is therefore important to set them out and assess them clearly and transparently.

Feasibility and Implementation

Feasibility analysis aims to establish for each of the options proposed whether it is capable of being implemented or is likely to come to a halt due to various impediments. Feasibility has several dimensions:

- Financial viability is the best known – it requires accounting for financial resources and the returns they may be expected to produce, and it may involve external sources and the partners to the project, although the partners may have different ways of measuring acceptable returns (Lichfield, 1998).
- Economic feasibility analysis has regard to the presence of sufficient demand to support the planned activities (for example, sufficient interest in certain cultural activities to sustain the new hall or commercial demand for the new employment floor space).
- Development feasibility analysis considers physical aspects – whether decent access, drainage, utilities, etc. can be provided, and it assesses the likelihood of obtaining planning consent and the likely co-operation of the owners of property (title) and other legal constraints.
- Public feasibility analysis considers the social and the political response, and attempts to anticipate where opposition may come from; such opposition can be reduced if the project management ensure public participation from the outset.

Any other impediments may be anticipated by tracing the process of implementation and the daily operations that follow. This approach also enables the project planners to identify the bodies that need to take part in the process. To the extent that these are not already within the partnership, contacts can be made to ensure understanding and collaboration.

Implementation will normally take place in stages. Each stage will incur costs and some may bring in revenues. Planning the stages can therefore be crucial for the viability of the project and advice should be sought from an expert in financial or development appraisal (subject to the contents of the project).

Impact Assessment and Comparison of Options

Although detailed project options are proposed with the intention of achieving the desirable objectives, the action is likely to have additional effects (traffic, demographic, environmental, etc.) which were not 'objectives' in themselves and may be beneficial or adverse to different groups of people (for example, a rise in house values is good for house owners but bad for tenants and purchasers).

The options should therefore be compared on grounds of impacts on all groups of the wider community, as well as on grounds of feasibility. Moreover, the notion of value for money is widely understood to mean not only financial and wider economic value but also social and environmental values. This is endorsed by the present administration, which advocates a holistic approach to regeneration, aiming to secure not only environmental sustainability but also a socially and economically sustainable society. By its nature this approach requires the coming together of minds and hearts of people from different backgrounds and concerns.

Conclusion

This takes the discussion back to the opening premises of this chapter: the trigger to success rests with the ability of the project manager to create a

---

Key Issues and Actions

- Address the questions of organisation and management at the outset of an urban regeneration scheme or project.
- Follow a clear defined pathway of organisation and management.
- Ensure that all actors and partners are aware of the arrangements.
- Keep good records.
- Manage, monitor and modify the strategy.
- Be willing to move to greater detail, but do not lose the strategic perspective.

work environment in which information, knowledge, concepts and sentiments are shared between all concerned. Although not all of those involved can sit on the partnership board, organisational structures and procedures can ensure their participation (in workshops, focus groups, committees, etc.) around particular issues in Phase 2, and around specific projects in Phase 3. This will enrich the project planner's understanding, contribute fresh ideas and mitigate potential opposition. Thus good management and good planning are intertwined.

References

Department of the Environment (1995) *Single Regeneration Budget Bidding Guidance*, Department of the Environment, London.

Lichfield, D. (1984) Alternative strategies for redistribution: changing approaches to renewal since the 1950s, *Habitat International*, Vol. 8, pp. 3–4.

Lichfield, D. (1986) Ashkelon's clean streets – face lift or body and soul treatment, *The Planner*, Vol. 72, p. 1.

Lichfield, D. (1992) *Urban Regeneration for the 1990s*, LPAC, London.

Lichfield, D. (1994a) Integrated regeneration strategies: looking behind the words, paper presented at a conference held at University College London, July.

Lichfield, D. (1994b) Assessing project impacts as though people mattered, *Planning*, 4 March.

Lichfield, D. (1997) Evaluation of urban regeneration projects – critique and proposals, paper presented at the UK Evaluation Society Conference, December.

Lichfield, D. (1998) Measuring the success of partnership endeavours, in *Public Private Partnerships for Local Economic Development*, Praeger.

Lichfield, D. and Lichfield, N. (1992) The integration of environmental assessment and development planning, Project Appraisal, September.

PART 4

EXPERIENCES ELSEWHERE AND A VIEW OF THE FUTURE

12   Lessons from America in the 1990s

John Shutt

Introduction

> While the great battles of the decades ahead are likely to be economic, the greatest challenge to our economic strength is certainly not competition from the Pacific Rim or Europe. No; the greatest challenge to our economic strength is here at home – where the decaying cores of too many inner-cities and the poverty-stricken heartlands of rural America threaten to erode our dynamic regional economies from within. That is what we intend to change. I believe we can do it.
>
> (Vice-President Al Gore, 6 December 1993, Boston, Massachusetts)

It is difficult to generalise about the United States and urban regeneration. The diversity and community enterprise which characterise its cities and regions are not always appreciated from a West European perspective. The USA and the United Kingdom have a long experience of transatlantic policy exchange (Hambleton and Taylor, 1993) and political parties, academics and policy-makers have regularly discussed policies for reviving urban economies with similar policy initiatives implemented both sides of the Atlantic (Clark, 1997b; Wiewel, Bennington and Geddes, 1992). Enterprise Zones, Community Development Corporations, Business Improvement Districts, Urban Development Corporations, public–private sector partnerships and coalitions for regeneration arouse similar concerns both sides of the Atlantic, but operate in wholly different contexts. Beneath the need for the interchange of ideas and best practice, it is the scale of the urban crisis in America which appears formidable to the European. There is little evidence of a strong coherent integrated federal state and local government approach to urban regeneration, which tackles the underlying structural issues forcing American cities into decline. The Clinton Empowerment Zone Policy for cities appears a weak tool when compared with the structural forces at work in American cities. It is perhaps, true to say that urban policy delivery and implementation in the United Kingdom is more coherent, centralised, planned and directed, whereas in the United States the scope for decentralised initiatives and community development and community capacity building is far greater.

This chapter considers:

- the major structural forces at work in urban America and the need for local government reform if the cities are to have an effective base;
- the Clinton approach to urban policy and the new 'Empowerment and Enterprising Communities Initiative' of the 1990s modelled on City Challenge;
- community economic development and Community Development Corporations, utilising a Washington case study;
- the development of public–private partnerships in cities, utilising case studies in New Orleans and Fort Worth;
- the likely future evolution of inner city economies in the USA.

Overview: Reviewing the USA Urban Crisis

Migration and Cities

Mass immigration is continuing to transform America. In the twenty-first century many United States cities will be non-white and black; Latinos and Asians will be half of the population. In 2015 whites will be in the minority in states like California, Arizona, New York, Nevada, New Jersey and Maryland (Mahindge, 1996). Much of the post-war period has been dominated by the movement of American whites out of urban America. Cities are already predominantly inhabited by black America, segregation is high and the turmoil and turbulence of city life is something for whites to avoid. The crisis of cities is widely perceived on both sides of the Atlantic as heading for chaos. In the United Kingdom and elsewhere in Europe, the need for Britain and Europe to develop new urban policies and avoid following the American urban 'nightmare' is a recurrent theme. This can be seen, for example, in the 2020 City Campaign mounted by the Labour Opposition Spokesperson for Local Government (Vaz, 1996).

The Economist (15 February 1997) reports in typical style:

> For many Americans the idea of California as the future is terrifying. Calamities such as the Los Angeles riots and the O J Simpson verdict have projected an image of a racially riven society, shivering on the edge of disintegration. Californians are increasingly living in ethnic enclaves, with whites retreating behind walls and Latinos recreating in their native barrios north of the border. A third of new housing developments in Los Angeles in 1990–1995 had locked gates.

In the same week, the London *Evening Standard* reported on the nightmare of Washington urban life (*Evening Standard*, 25 February 1997) confirming the 'British' view that the American inner cities are places to avoid if at all possible. Many urbanists, however, are aware that policy transfer between the USA and Britain is increasing and that Labour's intention of introducing a mayoral competition into the UK (Blair, 1998) is indicative of the strong policy interchange developing between the British Labour Party and the US Democratic Party. This policy exchange extends to interest in 'zone' programmes.

Industrial Restructuring and Edge Cities

Industrial restructuring in the United States, the changing competitive environment and new technologies and the globalisation of the US economy have transformed American cities and regions in the 1990s. White flight has created Garreau's 'Edge Cities' (Garreau, 1997) and the new 'CitiStates' (Pierce, Johnson and Hall, 1993) – suburbs with no single urban centre. Edge cities are 'independent' of the old central cities. Whites and aspiring ethnic minorities can function within the new edge cities without reference to the old central cities. In some urban systems the central business district may still be a functioning economic reality – in others it is having to find a new role – frequently a major tourist role, as economic and social activities have relocated outside cities. Outside the central business cores, the 'inner cities' are reserved for poor immigrants and poor whites. Inner cities can be full of entrepreneurship and community vitality but they can also be localities of despair, especially in a society where welfare benefits and local state expenditures are dwindling and subject to declining resource availability and radical restructuring of delivery systems.

Community Development

Community organisation and community development in the United States is the positive side of retreating local, state and federal government and centre urban distress. Dustin (1995) argues that

> community based organisations, more so than local government, symbolise something valuable worth taking into the 21st Century for many citizens and policy-makers. In retreat federal and state policy-makers proclaim the victory of free people, free-enterprise, civic responsibility and self reliance. In defeat they honor hallowed heroes – volunteers and bootstrap capitalists. The words truly inspire and strike a chord deep in the American soul. Empower citizens, build community, maximise local control, let the people decide, help people help themselves, count community assets, increase individual capacity, the elocution of these words evokes a feeling of doing the right thing. The vision seems to be this – sometime in the future, somehow and somewhere one person – one block – one city – at a time an alternative urban economy and governance will emerge which will be as productive as the industrial city and as inventive as the corporate city government once were.
>
> (Dustin, 1995, p. 2)

At neighbourhood level community development in the United States is spearheading the drive to decentralisation and the nation excels in its community-building capacity and its tradition of community organising. Community Development Corporations focusing on economic regeneration and revitalisation exist in their hundreds, providing inner city residents with a range of services that governments do not provide. The strength of community achievement against the backdrop of massive structural change and federal and state withdrawal is regularly what impresses most visiting academics and policy-makers. Neighbourhood vision and community involvement is impressive, but whether or not fragmented communities can combat the structural disparities between central cities and suburbs without a stronger federal urban policy is the central issue for many urban policy-makers. Americans have been preoccupied with reducing taxes and the role of federal government for much of the last decade and it is the cities which have suffered.

Federal Urban Policy

Urban policies in the United States were not the main concerns of the Reagan and Bush administrations in the 1980s. Reducing the influence of federal government was the preoccupation and Barnes (1990, p. 564) records how the federal government 'simply stopped or decreased doing certain things'. Federal government turned its back on the large USA cities with the Republicans preferring to support state governments rather than city governments. President Bush, addressing the National Urban League in August 1989, summarised the state of Urban America at the end of the 1980s thus: 'In many respects – let's face it – urban America offers a bleak picture – an inner city in crisis. And there is too much crime, too much crack. Too many dropouts, too much despair, too little economic growth, too little advancement – and the bottom line, too little hope' (The White House, 1989: quoted in Barnes, 1990, pp. 562).

The nation that had launched the War on Poverty and Model Cities programmes in the 1960s floundered throughout the 1980s as economic and industrial restructuring and federal withdrawal increased poverty and social deterioration in American inner cities (Davis, 1995). This deterioration was symbolised by the 1992 Los Angeles riots. The shock of the Los Angeles riots, arguably, began to force attention back on to the American inner city and the programmes required for regeneration. Within weeks the Washington-based Urban Institute was arguing for a new integrated approach to urban policy, addressing city problems within the context of an overall domestic policy stressing human investment, job creation and the breaking down of race and income barriers within the centre city: 'Traditional opportunity channels in the cities – manufacturing jobs, government jobs, good schooling – are shutting down. As a result, the American city, seen throughout much of its history as an engine of social and economic mobility without parallel in the world, is becoming a machine that reinforces inequality' (Urban Institute, 1992, p. 1).

Reinventing Urban Policy: a New Blueprint

Enterprise Zones were the dominant inner city regeneration idea first proposed by Jack F. Kemp in 1980 calling for legislation from the Reagan administration. The Thatcher government in the United Kingdom implemented the Enterprise Zone idea before Kemp in the USA. Kemp became Secretary of Housing and Urban Development under Bush and following the Los Angeles riots new legislation was brought forward by Kemp, but vetoed by Bush, in the run up to the 1993 US presidential election campaign.

Democrats were busy developing the Enterprise Zone idea. During 1992 Bill Clinton announced that he agreed with Kemp about the need for Enterprise Zones, but 'I think it's a very narrow view of what needs to be done. They will be of limited impact unless you also have the national initiatives I've called for on education, health care and the economy' (Lemann, 1994, p. 1). After the election, the Clinton administration revised the Republicans' abandoned enterprise zones bill, transferring it into the Empowerment Zones and Enterprising Communities policy and programme. Vice-President Al Gore became leader of the Enterprise Zone and Empowerment Zone initiative and Housing and Urban Development (HUD) Secretary Henry Cisneros – San Antonio's first Hispanic mayor in the 1980s – became responsible for pursuing Clinton's new urban policy programme.

The Empowerment Zone policy is one Democratic policy which has aroused recent interest in the United Kingdom, together with the role of US Economic Development Corporations, and US elected mayors, Community Development Corporations and Business Empowerment Districts, but as Clark points out US urban problems originate from a very different system of fragmented government and fiscal jurisdictions (Clark, 1997a). In transferring policy ideas we need to understand the very different context within which American cities have to operate. This different context exerts considerable influence on the urban regeneration process.

Empowerment Zones and Enterprise Communities

President Clinton signed the Omnibus Budget Reconciliation Act of 1993 (HR 2264) on 10 August 1993. The Bill made provisions for the designation of nine 'Empowerment Zones' (six urban designated by the HUD Secretary and three rural ones under the Agriculture Secretary). Designation of the zones was to be made by competition during 1994. In addition there were to be 95 'Enterprising' communities designated, 65 urban and 30 rural.

Empowerment Zones (see Table 12.1) had to be nominated by both local government and the state government, and an application for designation had to include a strategic plan co-ordinating activities which addressed:

- economic development
- human development

- community development
- physical development.

Plans had to bid for funding under federal programmes but also to identify the extent to which poor persons and families will be empowered to become economically self-sufficient. Empowerment Zones in the cities were to be urban areas of less than 200,000 people and in rural ones must have a population of less than 30,000. In any event they were to be areas of 'pervasive poverty, unemployment and general distress' (HR 2264).

Table 12.1   **Empowerment Zone Partnerships in the USA**

Federal level	State level	Community level
• Remove regulatory barriers	• Invest state resources and Federal funds provided to the state	• Involve the entire community
• Simplify programme rules	• Pass through EZ/EC SSBG funds	• Plan comprehensively
• Co-ordinate programme	• Co-ordinate programme and agencies	• Leverage private resources
• Invest broad resources	• Allocate a portion of private activity bond capital	• Streamline local government

Source: Community Enterprise Board (1994).

The New Grant Regime

Empowerment Zone status brought extra federal monies of $10 million per year for ten years for:

- Social Services Grants:
    - to prevent and remedy child abuse and neglect;
    - to achieve self-sufficiency (training and self-employment);
    - to achieve and maintain self-support (community and economic development);
    - to provide emergency shelter;
    - to support home-ownership programmes;
    - to support child-care institutions;
- Employment Credits – a tax credit for wages paid to 'zone employees' up to ten years;
- Tax Exempt Facility Bonds – bonds states and local governments could issue where 95 per cent of the net proceeds are used to provide an Empowerment Zone facility (property, land, infrastructure).

Competition and Urban Policy

Thus the Empowerment Zones idea differs from the enterprise zone concept in its focus on people rather than the search for inward investment and new businesses. The Clinton administration established the Community Enterprise Board (CEB), chaired by Vice-President Gore, and by the beginning of 1994 this was ready to formally launch the Empowerment Zone (CEB, 1994) initiative. American communities had six months to respond to the strategic guidelines and application process. The competitive bidding element of the programme and relatively modest resources left some observers sceptical from the start. Gramlich argued Empowerment Zones were a Grand Illusion dependent 'entirely on the determination, strength, and good luck of low income community groups. There are a few modest opportunities available but most groups will have to struggle to be included in the planning and implementation of zone programmes. Some groups will, rightly, judge the opportunities insufficient for the effort' (Gramlich, 1993, p. 2)

The Clinton administration, however, was clear that the Empowerment Zone legislation built on the anti-poverty programmes of the 1960s and made a better case for community development in the context of a new partnership of stakeholders, working together in key designated neighbourhoods. 'Communities that stand together are communities that can rise together. Communities cannot succeed with public resources alone. Private and non-profit support and involvement are critical to the success of a community seeking revitalisation.' (CEB, 1994, p. 1).

The Winning Cities

In December 1994, the successful Empowerment Zones were declared in six cities after a national competition and rigorous evaluation by HUD and vigorous political lobbying. Atlanta, Baltimore, Chicago, Detroit, New York and Philadelphia/Camden were the winners, and the pressures to enlarge the number of areas to benefit led to Los Angeles and Cleveland being declared Supplemental Empowerment Zones (SPZs).

Evaluating Partnership

Gittell (1996a; 1996b) evaluated the progress of the zones throughout their first year and examined community involvement in the bidding process. She concludes that the first stage of the process did little to enhance the development of civic community capacity, although the established Community Development Corporations in Detroit and Chicago did play strong roles in the process. Mayors and institutional actors played the key roles and made sure their favoured projects were designated inside the zones. She records that business involvement has largely been limited to banks with a direct interest in the Empowerment Zone neighbourhoods.

Coca Cola and Turner Broadcasting are involved in Atlanta, Campbell Soup has a strong presence in Camden, General Motors in Detroit, but all of their roles are historical and traditional as major actors in the city rather than as active and engaged entrepreneurs seeking new and significant investment in the Zones.

(Gittell, 1996b, p. 40)

Modares (1996), researching the Los Angeles Supplemental Empowerment Zone, questions the concentration of the Clinton administration's Empowerment Zones on the East Coast and argues against an Empowerment Zone methodology which supports area-based poverty for specific types of communities and excludes low-density areas like Los Angeles. Moreover, the Supplemental Empowerment Zones lack the social service grants of the prime six Empowerment Zones, which means that the focus is on local business growth industries and attracting businesses rather than on community and social development approaches.

Mayor's Prerogative

The Empowerment Zone process itself has focused on the favoured projects of mayors who often went out of their way to make sure that the boundaries of Empowerment Zones included key projects and initiatives. Again Gittell points out that

> Atlanta Mayor Cambell made certain that the downtown neighbourhood which housed the Olympic Stadium was designated as part of the Zone. In Detroit, General Motors' new plant was a priority for inclusion in the Zone and the mayor made it happen. In Baltimore, the John Hopkins Complex, a major contributor to the city's economy, received priority from Mayor Schmoke and was included in the boundaries of the Zone. Republican Rudolph Guiliani saw only one purpose to the programme, some guarantee that it would provide funds for Yankee Stadium so that the team would not leave New York City.
>
> (Gittell, 1996b, p. 39)

Second Term

In his second term, President Clinton is concentrating on developing a second round of the Empowerment Zone concept. In 1996 it was announced that 'home-ownership zones' modelled after the Enterprise Zone concept would also be a Clinton administration priority alongside 'Empowerment Zones' and that the second round of awarding Enterprise and Empowerment Zones, round two, would now be initiated (*National Mortgage News*, 26 February 1996). Cisneros also claimed that the Empowerment Zones were showing signs of success with Detroit attracting $2 billion of investment into the city as a result of Empowerment Zone creation.

The similarity with the British City Challenge and Single Regeneration Budget 'competition' has been observed by Hambleton (1996). What is interesting about the Empowerment Zones, however, is the greater focus on strategic planning, and community participation, and the linking of federal, state and city actors, and the different basis of the programme approach compared with the United Kingdom. Practitioners in the UK need to follow the progress of the Empowerment Zone concept and its results with more interest throughout the 1990s, particularly since a new Blair Labour administration in the UK is making great use of the zone concept for health, education and employment zones. Following the designations, little is known about the impact on the losing cities, although Schulgasser (1995) suggests that it has forced a city like Newark to start to think about strategic planning again, and this implies meeting regularly with external groups.

The Empowerment Zone process signalled the start of the regeneration of HUD, whose mission 'places it squarely in the center of poverty and racial realities that do not lend themselves to simple answers' (US Department of Housing and Urban Development, 1994). Under Clinton's radical Reinvention Blueprint, HUD was required to change from being a top-down, bureaucratic, complex and overly prescriptive department to becoming a true partner in communities – acting as a clearing house for innovative solutions. Meyer (1995, p. 23) praises HUD for raising the level of discussion and understanding on the Empowerment Zone/Enterprise Zone programme and in particular for its attention in the new economic development programme. Housing and Urban Development understands the need to couple job training, job placement and linkage and supportive services with business assistance and related job-generation approaches. Better jobs targeting for inner city residents is seen as the real promise of the Empowerment Zone initiative and HUD is seeking to support the Empowerment Zone/Enterprise Community initiative by incorporating other federal resources and programmes.

In the United Kingdom in 1996, Blunkett, Shadow Education Secretary, proposed a new set of 'Training and Employment Zones' for British inner cities, designed to achieve a similar focus on jobs targeting and the inner city residents (Blunkett, 1996). This illustrates once again the way in which US and UK urban policies appear to be moving closer together in similar directions under respective political parties. There is constant interaction and interchange between politicians, academics and policy-makers leading to the implementation of similar initiatives and programmes in terms of national urban regeneration policies. Competitive bidding and the zone mentality have become the dominant urban policy delivery process, both sides of the Atlantic.

Community Capacity and Community Development Corporations

Community Development Corporations 'bridge the gap between capitalism and community development' (Harrison and Weiss, 1993). They are interesting self-help initiatives often delivering services to black and hispanic communities which are of interest to those in the UK who would like to see a new focus on black economic development and community development in the late 1990s. Since the 1960s they have become a potent force in American urban revitalisation. Community Development Corporations (CDCs) started with an affordable housing and community development remit and have gradually extended this role. Non-profit corporations, CDCs have developed a job-creation and economic development role and extended their activities into job training, placement and counselling activities. This extension of the role was encouraged by successive federal programmes in the 1980s, like the Job Training Partnership Act (JTPA) or the Perkins Vocational Education Act. Community Development Corporations like the Newark Community Corporation (NCC) in Newark began to review their own job-training programmes in the 1990s. Spurred on by foundations like the Urban Poverty Programme of the Ford Foundation, and the Enterprise Institute, nationally known for its affordable housing programmes, CDCs have been assisted in developing employment and training projects (Harrison and Weiss, 1993). Housing and Urban Development under Clinton has refocused its efforts on encouraging Community Development Corporations and put more of their efforts into economic development and enterprise initiatives. As the CDCs have developed so they have developed their social service provision and elderly service provision and into youth services, providing a role for example, in Aids education and drug substance abuse prevention. The CDCs frequently provide services now in inner city areas, which in the United Kingdom are currently still provided by central or local government, the National Health Service, or other quangos. Increasingly as the welfare state is reduced and public expenditure programmes are reduced throughout Western Europe, CDC models are being developed and examined in the UK.

Washington, DC, Case Study

Community Development Corporations have been described in the UK as 'housing associations with knobs on'. In the UK the Community Development Trusts Association is promoting the CDC idea and local authorities like Brighton and Liverpool are examining the policy framework for developing CDC approaches to neighbourhood development.

In Washington the Marshall Heights Community Development Organisation (MHCDO) provides a case study of good practice in CDC development. In the mid- 1990s Washington's District Council had a major $300 million structural deficit and the Commission of Social Services reported in 1994 that:

- nearly half of the district's children are on welfare;
- one out of every eight residents receives welfare;
- one out of every six residents is on food stamps;
- one out of every five residents receives Medicaid.

And the sad truth is that the city has no realistic way of keeping up with these spending demands. Despite the urban town seen by tourists when they visit the nation's capital, the city away from the mall has the woes and responsibilities of a state and the purse of a hamlet.

(The Washington Post, 8 March 1994).

Washington like other capital cities has a declining population as affluent households continued to move out through the 1970s and 1980s leaving the district for the suburbs. Businesses followed and inner city neighbourhoods like Anacostia found that they were becoming unviable. The MHCDC began back in 1979, becoming the co-owner of the East River Park Shopping Centre and initiating major renovation, developing housing, a small industrial park as a business incubator facility and, in the 1990s, developing its drug rehabilitation and social service programmes.

Whilst the MHCDC operates in a deteriorating Washington environment, and its achievements can be seen as small-scale, the organisation has provided a positive neighbourhood driving force for revitalisation. The MHCDO works in Wards 7 and 6 East neighbourhoods which experience the highest levels of documented illegal drug use, associated violence and youth homicides than any other sections of Washington. The MHCDO has evolved into a multifaceted organisation with a $2.5 million budget (1994). It has been commended by the Local Initiatives Support Corporation (LISC) as a 'model for rebuilding troubled cities from the grassroots' (MHCDO, 1994).

The MHCDO sees itself

- creating new and diverse economic opportunities;
- mobilising and empowering residents to become involved in the process;
- encouraging a holistic approach to community development;
- assisting in the development of CDCs;
- assisting community-based groups to develop the four Cs: capacity, collateral, credit and character.

Local Economic Development and Regeneration Partnerships

Community economic development initiatives in the USA are less well known than the growth and urban development coalitions which received a great deal of academic attention in the 1980s. Physical developments 'showy brick and mortar developments, complete with ribbon cutting ceremonies' (Swanstrom, 1985) won the admiration of the Thatcher administration in cities like Baltimore and Boston and provided added impetus to the British Urban Development Corporation programme, especially those

*Box 12.1   Marshall Heights Community Development Organisation, Inc.
Washington, DC: history and achievements*

The Marshall Heights Community Development Organisation, Inc is a not-for-profit, community-based organisation that was formed in 1979 by a spirited and insightful group of residents from the Marshall Heights neighbourhood. Shortly thereafter, the group expanded its membership to include the neighbourhoods of Deanwood and Burrville.

The corporation was created to provide an organised forum for citizens to express community concerns on housing and community development and to facilitate citizen interaction with the District of Columbia Government.

Summary of Achievements, 1979–94

- With a $25,000 grant for venture capital from DHCD, MHCDO leveraged a $3.2 million deal for the purchase and exterior renovations of the East River Park Shopping Center, located at Minnesota Avenue and Benning Road, NE. The MHCDO is both an equity owner of the property and a co-general managing partner. The renovations included a new façade, new parking lot, major landscaping and additional lighting. Now, the community is better served and the shopping centre continues to turn a profit under community leadership.
- The MHCDO, as a participant in the HomeSight Program, a co-operative effort among local community development corporations, the Local Initiatives Support Corporation, Fannie Mae Foundation, the DC government, and participating lenders, is continuing to develop affordable housing for sale to buyers who qualify for assistance under the City's Home Purchase Assistance Program (HPAP). Through its subsidiary, Citizens Housing Development Corporation, a total of 23 houses have been rehabilitated and sold to HPAP clients by 1995.
- In an expansion of the HomeSight housing development programme, Citizens Housing Development Corporation (CHDC) began infill development of new modular homes in 1990, inaugurating the programme with four new homes at Drake Place, SE. The street has since been renamed 'Queens Stroll Place' to commemorate the visit by Queen Elizabeth to these new homes in May of 1991. On 4 August 1993, CHDC broke ground for the Deanwood Station townhouse style condominiums opposite the Deanwood Metro Station. The project consists of 12 homes and is the largest single new construction venture undertaken by CHDC to date.
- Multi-Family Housing also includes Magnolia Gardens, a 13-unit building; 1449 Olive Street, NE, a four-unit building; 3426 and 3421 Minnesota Ave, and 3001 Nelson Place, SE renovated as condominiums.
- The Kenilworth Industrial Park project features 90,000 square feet of space for light industrial use on Kenilworth Avenue, a business incubator for start-up businesses, as well as manufacturing and commercial space. 1235 Kenilworth Avenue, NE, a 13,000 square foot multi-story structure, is operational as the first Business Incubator facility in Washington, DC, providing below market rate space to small start-up commercial tenants.
- In March 1990, the Marshall Heights Community Development Organization received a grant of $100,000 from the Robert Wood Johnson Foundation (RWJF), in Princeton, New Jersey, to plan a comprehensive drug and alcohol prevention, reduction and treatment programme within the Ward

7/6 East of the River community. In April of 1992, the RWJF awarded MHCDO a one-year grant of $599,916 to implement the 'Fighting Back' Program. The national RWJF initiative, 'Fighting Back', is designed to create both long-term prevention and treatment that reduces the demand for drugs and alcohol in carefully targeted neighbourhoods. In February 1993, the MHCDO was awarded an additional 18 months of funding for a total grant amount of $3 million from RWJ.

• The MHCDO also offers social service programmes under contract to the Federal Government that include the Michaux Senior Center, which provides activities and educational programmes for Ward 7 seniors; Employment Training and Development and Black Male Employment Training programme which targets black males between 18 and 25 years of age which utilises classroom training and career assessment in preparation for participants to successfully compete in the labour market; and a Crisis Intervention programme that provides emergency food and clothing to residents in crisis situations.

Source: MHCDO (1994).

faced with port regeneration projects. The study of urban growth regimes has been important in the USA (Logan, and Molotch, 1987; Stone, 1989) with governing coalitions in cities focusing on mayoral leadership of American cities alongside local business interests in economic and land development partnerships. Strong mayoral leadership is seen as critical in providing a focus for American economic development policy (Wolman and Spitzley, 1996).

The Economic Regeneration Partnerships and Economic Development Corporations have been critical factors in successful economic development and regeneration strategies. Boston in the 1980s under Mayor Flynn is frequently held up as an example of successful partnership development between state and local government, business and the philanthropic community, and with community developers and community organisations. Boston's job policy and the Boston Compact (training programmes) and Boston's housing policy have received much attention (Drier, 1997). Boston's experience of economic growth and a declining poverty rate and strong local government led action is, however, atypical of most of the major American cities.

New Orleans

The city of New Orleans is more typical of American cities facing severe economic difficulties where 'out-migration continues to be one of the largest obstacles to economic development' (Barthelemy, 1992). See Table 12.2 for further details of population change.

The New Orleans economy was tied to the oil and gas industries and boomed throughout the 1970s. In 1975 the Superdome was built, home of the New Orleans Saints football team, and in 1985, the city opened the New Orleans Convention Centre which made New Orleans one of the top convention centres in the United States. In the early 1980s, the city created the Almonaster-Michoud Industrial District (AMID), which is said to be the largest industrial park within the boundaries of a single city in the United States. The AMID is home to the Martin Marietta's production operations for the external fuel tank for the NASA space shuttle. In the 1990s the city began to redevelop the Mississippi Riverfront area downtown. It constructed the Aquarium of the Americas which opened in 1990 and embarked upon further riverfront tourism development, which is taking place throughout the 1990s (see Box 12.2).

Despite a booming service and tourism sector in the Central Business District and French Quarter of the city, the city suffers from employment decline in most sectors of the city's economy (Tables 12.3 and 12.4) and because the skilled labour pool is depleted by out-migration, critical labour shortages of trained workers have occurred in key sectors, e.g. in health care professions. The 1992 Economic Development Strategic Plan for New Orleans put forward the following major obstacles to economic development in the city in the 1990s:

- a relatively poorly educated population;
- a poorly funded local public education system;
- a high proportion of its population below the poverty level;
- lack of public funds at local, state and federal levels;
- lost tax revenues to the Louisiana State government caused by the decline in oil prices and its impact on tax revenue for the city of New Orleans;
- overdependence on business tax and mineral revenues creating a tax climate viewed as being onerous to business.

The perception that economic development activities had not been well co-ordinated between public and private sectors saw the creation of a regional Metrovision Partnership in 1990 to create a new economic development agenda for the twenty-first century. Metrovision was a process

Table 12.2 **Population characteristics of New Orleans**

(a) *Population in the city of New Orleans and its suburbs 1960–90*

	1960	1990	% Change
Central city	627,525	496,938	−20.8
Suburban counties	240,955	741,278	+207.6
New Orleans metro	868,480	1,238,216	+42.6

(b) *Race and ethnicity in New Orleans 1960–90*

	1960	1990
White	62.6%	33.1%
Black	37.2%	61.4%
Other	0.2%	2.0%
Hispanic	—	3.5%

Source: City of New Orleans, 1994. See also Lauria, M., Wheelan, R. and Young, A. A., (1993)

Table 12.3   **Net employment changes in New Orleans Parish, 1980–89**

	%
Agriculture	–11.8
Mining	+12.2
Construction	–17.2
Manufacturing	–47.6
Transport	–33.1
Wholesale trade	–40.4
Retail trade	–11.7
Finance/insurance/real estate	–9.0
Services	+13.0
Government	–7.8

Source: Louisiana Department of Employment and Training Annual.

Table 12.4   **Economic impact on New Orleans Parish from tourism, selected years 1982–92**

	Jobs	Payroll	Revenue local taxes
1982	39,584	$539,821,700	$35,137,400
1985	40,048	$512,229,700	$40,121,700
1987	46,196	$580,962,300	$49,065,600
1989	40,840	$498,512,500	$45,931,300
1990	45,320	$524,131,100	$51,664,932

Note: Dollar amounts adjusted using 1991 GNP Deflator
Source: University of New Orleans Office of Business and Economic Research; Travel Impact Model

of bringing together key public and private stakeholders in the region to develop a strategic plan. 'Making our region competitive is the message and mission of the Metrovision Partnership' (Metrovision, 1990). Following the Metrovision initiative the city of New Orleans developed a new Strategic Planning Committee and completed a new Economic Development Plan for the city of New Orleans during the tenure of Mayor Sidney Barthelemy. Beginning in May 1991 action plans were formulated for:

- warehousing, distribution and port regeneration;
- image improvement;
- arts, sports and music development;
- technology-based industries;
- downtown office/medical centre development;
- job training;
- neighbourhood revitalisation/housing;
- small and minority business development.

(see city of New Orleans, 1992)

Under the leadership of local business interests and community leaders the city embarked on a major programme of tourism development and marketing, building on the prospects of increased growth in the convention business, despite the evidence of tourism's seasonality, low pay and transport infrastructure demands.

Whether or not a tourist development strategy can act as the backbone of economic regeneration in a city is a question which is frequently being asked in the United Kingdom. However, New Orleans has moved its economy to the point where it is one of the most favoured tourist destinations in the United States, although the impact on its inner city localities remains limited. Success in economic regeneration and physical regeneration in the central business cores of American cities does not ensure successful trickle-down and aid to the inner city residents without sustained efforts to develop linkage programmes. Many of the service jobs which are created are low wage and low skilled. Even with its new sense of public–private partnerships and strategic economic development planning, New Orleans was not successful in winning resources through the first round of the empowerment zone competition in 1994/95.

Box 12.2   New Orleans Riverfront 2000: Overview

The New Orleans Riverfront 2000 plan is the largest development project in Louisiana's history, and by the turn of the century will make New Orleans one of the most exciting riverfront cities in the world.

New Orleans Riverfront 2000 includes the construction of seven separate facilities along the Mississippi riverfront over the next ten years.

New Orleans Riverfront 2000 calls for the additional expansion of the New Orleans Convention Center, which would make the facility one of the largest exhibition halls in the world with over 1 million square feet of contiguous exhibition space.

New Orleans Riverfront 2000 includes the expansion of the Riverfront Park from the Moonwalk to the Mandeville Street Wharf with the park acting as the gateway to the city and the front door to the historic French Quarter.

New Orleans Riverfront 2000 involves the construction of a conservatory, which will display rare and exotic plants, to be located within the Riverfront Park expansion and managed by the Audubon Institute. This concept has been altered to an insectarium which is now under negotiations for placement in the Custom House, an important historic structure which is owned by the Federal government, near the riverfront end of Canal Street, the city's main downtown street.

New Orleans Riverfront 2000 includes the construction of Phase II of the Aquarium of the Americas adding more than 60,000 square feet for new underwater exhibits.

New Orleans Riverfront 2000 will also include the construction of a $40 million natural history museum, which would be built next to the conservatory and would be another living science facility managed by the Audubon Institute.

All seven projects in the Riverfront 2000 plan are projected to create more than 17,000 jobs and will generate more than $725 million a year for the local economy and raise more than $66 million a year in state and local taxes.

Riverfront 2000 will be funded by a combination of city and state bond issues and private donations raised by the Audubon Institute.

Riverfront 2000 will attract nearly 1 million new visitors to the Crescent City each year.

Riverfront 2000 will put New Orleans on the map as one of the fastest growing family destinations in the country.

New Orleans Riverfront 2000 includes the completion of a new Audubon Zoo master plan and the construction of the Species Survival Center for endangered animals.

New Orleans Riverfront 2000

	Project cost	*Opening*
Species Survival Center	$10 million	1992
Riverfront Park expansion	$15 million	1992
Conservatory (Changed to insectarium placed in existing building)	$10 million	1993
Aquarium of the Americas (Phase II)	$20 million	1994
Convention Center (Phase III)	$156 million	1995
Natural History Museum (later eliminated from plan)	$40 million	(not implemented)
Zoo 2000 Master Plan	$17 million	1997
Riverfront 2000 Total	$268 million	

Source: New Orleans Riverfront 2000 'Gateway to the Crescent City', December 1989.

New Orleans is a city struggling to come to terms with both the restructuring of the US economy and the polarised society this has created. On the one hand it is the city of crime, violence and urban distress (384 murders in New Orleans in 1994). On the other hand it is the city of carnivals, Mardi Gras and Jazz Fest and the city of the Garden District with its million-dollar houses, the French Quarter with its artists and literary eminences, and the birthplace of Jazz. Thirty per cent of the city's entire population live below the federal poverty level (Washington, 1995). It failed to be selected for Empowerment Zone status, but it was chosen to be one of the Enterprise Communities.

Despite urban distress one of the features of the American mayoral system is the scope for 'municipal enterprise' (Table 12.5). In order to attract hotel and sales taxes to boost city coffers, many US cities like New Orleans have been forced to develop and expand their convention centres and are in sharp competition with each other. San Francisco, Los Angeles, St Louis, Kansas City, Dallas, Salt Lake City, Sacramento, Baltimore, San

Table 12.5   **Top ten employers in Greater New Orleans, 1992–93**

Rank	Employer	Employees	Industry
1*	US Department of Defense	18,000	Military
2	Orleans Parish Public Schools	8,805	Education
3	City of New Orleans	7,874	Government
4+	Avondale Industries	7,100	Shipbuilding/construction
5	Tulane University	5,326	Education
6	Energy Corp	4,340	Utilities
7	Ochsner Medical	4,105	Health care
8	Schwegmann Giant Supermarkets	3,900	Grocery stores
9*	South Central Bell	3,048	Telecommunication
10*	Martin Marietta Manned Space Systems	3,000	Space shuttle external tank

Note: * Company not headquartered in New Orleans.
        + Fortune 500 Company headquartered in New Orleans.
Source: New Orleans City Business 1992–93; Orleans Parish School Board; New Orleans Civil Service Board.

Diego, Orlando and Chicago have all been expanding their convention centre economic development strategies in the 1990s (Nelson, 1996). Competition on this scale between cities is much greater in the USA than in the UK.

Fort Worth

The use of cultural industries strategies to boost tourism and urban regeneration is very much in evidence in Fort Worth, Texas, (see Box 12.3), sponsored by the drive of one family – the Bass family – and buttressed by the sponsorship and donations of many influential foundations which include the Richardson Foundation, the Burnett-Tandy Foundation and the Carter Foundation. Their influence can be compared to the nineteenth-century role of the UK's Victorian entrepreneurs – Cadburys in Birmingham or Pilkingtons in St Helens – but in a modern-day context and raising important issues of city ownership, power and influence in the urban regeneration process. This is a city benefiting the growth of the US economy in the late 1990s without the levels of urban distress of the kind evident in Washington and New Orleans.

Conclusion

In the 1990s competition dominates urban policy and economic development policy on both sides of the Atlantic. For Al Gore 'empowerment zones and enterprising communities are perhaps the most vivid examples of President Clinton's new approach to rebuilding neighbourhoods, creating jobs and changing lives' (Gore, 1996).

Box 12.3 Cowboys and culture: the Fort Worth Partnership, Forth Worth City

'We wouldn't be telling it to you straight if we didn't brag about our cowboys. Our Chisholm Trail. Our historic stockyards, where the steaks came bigger than the platters. Or our fancy fiddlin' and soft country ballads at Billy Bob's. That's Fort Worth' (Fort Worth, Catching the World's Attention, 1997).

The renaissance of downtown Fort Worth is one of the USA's central city success stories of the 1990s. With its historic downtown area based on the successful urban regeneration of Sundance Square and its Cultural District boasting world class art museums and the largest science and history museum in the south-west, it is a city which is utilising the cultural industries and tourism as the springboard for regeneration. The Kimbell Art Museum has been described as 'America's best small museum' designed by famed architect Louis I. Khan. The Stockyards National Historic District which houses the 1904 Livestock Exchange and embodies true Wild West cowboy livestock legacy is only 2 miles from the downtown centre. Besides the Stockyard Station, it incorporates Billy Bob's Texas 'the world's largest western dance hall'. The stockyard closed in the 1960s and the restructuring of the livestock industry impacted on downtown Fort Worth.

The renaissance of Forth Worth Downtown is dominated by the Bass family – Nancy Lee, Perry R. Bass and their four sons – who are widely credited with changing the face of Forth Worth. Their mission has been to bring new housing and commercial activity into Sundance Square and to ensure citizen safety by developing their own private security police force to patrol the city streets and successfully reduce crime. Youthful police on bicycles can be seen at all times on the city streets providing welcome reassurance to tourists and residents alike. The Bass family fortune was originally based on oil. Sid Bass as leader of Bass Brothers Enterprises built the family fortune through the 1960s and 1970s on property development and the regeneration of Fort Worth. Sid Bass built the Worthington Hotel and Sundance Square; brother Ed Bass developed the Caravan of Dreams Jazz Clubs, Sundance West apartments and movie theatre and has spearheaded the development of the $65 million Nancy Lee and Perry R. Bass Performance Hall which opened in May 1998 with 2,056 seats. Bob Bass created Cook Children's Medical Centre and is Chairman of the National Trust for Historic Preservation. Lee Bass, the youngest son, led the fight to refurbish the Fort Worth Zoo. Ranked thirty-seventh in the *Forbes* magazine list of the world's richest people in 1997, the Bass family worth is estimated at $6 billion and their contribution to regeneration is supplemented by the Sid Richardson Foundation, which supports education, health, human services and the arts in Texas.

'The Basses could have invested and do invest, all over the world. But they have chosen to invest here in Fort Worth in a very meaningful way, and the community recognises it and appreciates it' (Forth Worth Mayor Kenneth Barr, *Fort Worth Star-Telegram*, 26 April 1998).

The scale of the private sector contribution to economic regeneration supplemented by substantial foundation funding – the Carter Foundation fuels activity in the Fort Worth Cultural District – is what distinguishes the USA regeneration process compared to the small-scale foundation funding available in British cities.

Fort Worth is benefiting from the enormous growth of small metroplex edge cities in North West Tenant County for example, based on the cities of Lakeside, Lake Worth, Reno, Riveroaks, Saginaw, Sanctuary, Springtown, Westworth Village and Azle. This Texan Westplex Alliance of Communities rests on the growth of the Alliance Corridor and Alliance Airport developed by Ross Perot. Here a new communications and transport centre has been created based on major inward investment: Fedex, Intel, Motorola, Mitsubishi and Nokia, Southwestern Bell, Tandy Corporation and the home for corporate headquarters of IBM and Burlington Northern Santa Fe, bringing with it the incredible mobility and change which characterises Texas in the late 1990s. All of this is occurring some 40 miles north of Downtown Fort Worth. It is giving rise to predictions that Fort Worth will be bigger than the Dallas conurbation as a major metropolitan growth centre of the twenty-first century.

The Empowerment Zone programme has taken three years to operationalise and yet the policy itself is operating only in six major cities and on a smaller scale in the 95 'enterprising communities'. Whilst the Empowerment Zone policy was being forged and barely implemented, a Democratic President found his administration circumscribed by a Republican Congress and the progressive agenda for urban policy envisaged in the early 1990s began to disappear. The HUD's budget has been reduced from $26 billion to $19 billion and this has been accompanied by devolution to the states, welfare decentralisation and benefits cuts which threaten to further destabilise the inner cities and the low income poor beyond the deteriorating position of the 1980s. Throughout 1995 and 1996 a Republican Congress and Democratic President have pushed further on to the states responsibility for policy areas traditionally programmed and administered from Washington federal levels. In welfare and Medicaid, transportation, job training, environment, a Republican Congress and now Republican state governments are pushing costs further down to local governments. Inevitably the conditions in cities will deteriorate further as welfare programmes in particular are reduced. There are many pessimists in US urban policy who now believe that the development of federal intervention programmes is impossible to achieve (Meyer, 1995). In this situation local and regional initiatives in community economic development and regeneration become more important. However, the task of addressing the major structural issues of local government reform and fiscal reform of metropolitan areas remains to be achieved. Cisneros as Secretary of HUD continued to argue for regional reform of the fragmented local government system, albeit from a weak HUD base. 'People Regionalism', 'Cities without walls', 'We're all in it together' remained his theme tunes throughout the mid-1990s. Typically Cisneros argued that

> For those seeking opportunity today, the Mecca must be the entire metropolitan area – the central city and especially its suburbs. Achieving the American dream for everyone requires opening up all the metropolitan area's resources and opportunities to all its residents. Only with diversity, balance and stability everywhere can the decline of inner cities and aging suburbs be reversed.
>
> (Cisneros, 1995)

Fine words. However, Cisneros has already been replaced by Andrew Cuomo. There is concern about the future of the empowerment zones

---

Box 12.4 *Resources for Community Economic Development: the Center for Community Change, Washington*

The Center for Community Change (CCC) is a national non-profit organisation that works to revitalise poor communities in three ways:

- It provides free technical assistance and training to community-based organisations in low-income and minority neighbourhoods across the nation.
- It helps poor people achieve a voice on public policies that affect them.
- It launches special projects to conduct studies, test new approaches to community development and broaden support for worthwhile grassroots efforts.

At the Center for Community Change, staff believe the most effective way to improve conditions in low-income neighbourhoods is to help the people who live there empower themselves. To that end they help grassroots groups in low-income communities develop strong leadership, carry out effective strategies and create productive solutions to community problems.

The objective is not only to see small businesses established where no jobs exist, or crumbling tenements replaced with safe housing. The objective is also to see that grassroots groups develop the capacity to make these things happen for themselves. This means providing long-term, on-site assistance to urban and rural grassroots groups that address a daunting range of poverty-related issues.

The Center for Community Change pioneered this approach over 25 years ago. It still works. In 1994 CCC provided technical assistance to over 200 community groups comprised of African Americans, Asians, Latinos, whites, American Indians, farmers and farm workers, factory workers, female heads of households, immigrant families, street vendors, youth, the unemployed and others. To each of these groups the CCC provides tools which help them create the changes they envision in their own community.

The CCC's technical assistance staff includes experts in organisational development, community organising, action research, housing and economic development, community reinvestment, and other fields. All the Center's fieldwork is directed towards building the capacity of grassroots groups.

Headquartered in Washington, the CCC has a San Francisco office and staff outstationed in several other cities throughout the United States. The Center draws its support from more than 80 sources, including foundations, corporations, national churches, individuals, the federal government, and an endowment.

Source: Center for Community Change (1996).

---

programme initiative and behind this recognition that the USA's urban crisis and the future of American cities demands a far-reaching reorganisation of an outdated local government system not simply a zone completion. The local government system isolates poor cities within boundaries which can no longer provide the tax base to develop the services needed to regenerate communities. The rich white suburban communities use cities but do not pay for them. The need for local government reorganisation has never been greater – recognising the reality of modern competitive regional economies and labour markets, allowing for equalisation of fiscal resources and addressing strategic city-state policy issues such as transport planning, environmental sustainability, pollution control. However, politically such a reorganisation appears impossible to achieve in the United States that is emerging into the twenty-first century. There is little ground for confidence that state and local governments can work together to transform the structural problems of American cities. There is a concern that Clinton is already now immobilised and his policy initiatives have failed to make the major impact required at the city level. Although there can be little doubt that the booming economy has deflected the worst urban stresses from emerging, in the late 1990s, it is against this wider backdrop that the empowerment initiatives and urban regeneration programmes of the Clinton years will need to be assessed.

Note

I am grateful for the assistance of the Urban Affairs Association and the helpful comments of Andrew Mott, Center for Community Change, Washington; Jane Brooks, University of New Orleans and Peter Meyer, University of Louisville in preparing this chapter.

References

Barnes, W.R. (1990) Urban policies and urban impacts after Regan, *Urban Affairs Quarterly* Vol. 25, no. 562–573.

Barthelemy, S.J. (1989) New Orleans Riverfront 2000 'Gateway to the Crescent City', 20 December, Mayor's briefing paper.

Blair, T. (1998) *Leading the Way. A New Vision for Local Government*, IPPR, London.

Blunkett, D. (1996) Opportunities to earn, Labour's proposal to tackle long-term unemployment. Speech to CLES National Conference, November.

Center for Community Change (1996), *Annual Report*, CCC, Washington.

Cisneros, H.G. (1995) *Regionalism: The New Geography of Opportunity*, US Department of Housing and Urban Development, Washington, DC.

City of New Orleans (1992) A blueprint for economic revival. The Economic Development Strategic Plan for New Orleans, Mayor's Strategic Planning Committee.

City of New Orleans (1994) *A Statistical Chartbook 1960–1990*, College of Urban and Public Affairs, University of New Orleans.

Clark, G. (1997a) The Mayor's tool box, Local Economic Policy Unit, Southbank University paper, 1997.

Clark, G. (1997b) Transatlantic local lessons, *Local Government Chronicle*, 21 February.

Community Enterprise Board (CEB) (1994) *Building Communities Together: Empowerment Zones and Enterprise Communities Application Guide*. US Department of Housing and Urban Development, Washington, DC.

Davis, M. (1995) Who killed LA? A political autopsy, *New Left Review*.

Drier, P. (1997) The new politics of housing: how to rebuild the constituency for a progressive federal housing policy. *American Planning Association Journal*, 5–27.

Dustin, J.L. (1995) *Decentralisation in an Interdependent World*, Department of Affairs and Geography, Wright State University, Ohio.

The Economist (1997) 15 February.

Fort Worth Star-Telegram (1998) Unveiling Bass Hall, 26 April.

Garreau, J. (1997) *Edge City: Life on the New Frontier*, Doubleday, New York.

Gittell, M. (1996a) Expanding civic opportunity: the urban Empowerment Zones. Paper to the Annual Meeting of the Urban Affairs Association, New York City, 13–16 March.

Gittell, M. (1996b) *Community Empowerment and Empowerment Zones, Community Organisations as Stakeholders*, Graduate School, City of New York.

Gore, A. (1996) White House Fact Sheet: Empowerment Zones and Enterprise Communities.

Gramlich, E. (1993) *The Enterprise Zone Provisions in HR 2264*, Centre for Community Change, Washington, DC.

Hambleton R. (1996) Empowerment Zones and UK urban policy: competing in the urban regeneration game. Urban Affairs Association papers, March.

Hambleton, R. and Taylor, M. (eds) (1993) *People in Cities. A Transatlantic Policy Exchange*. School for Advanced Urban Studies, University of Bristol.

Harrison, B. and Weiss, M. (1993) Building bridges: Community Development Corporations and the world of employment training. Final report submitted to the Urban Poverty Programme of the Ford Foundation, Carnegie Mellon University, Pittsburgh, PA.

Lauria, M., Wheelan, R. and Young, A.A. (1993) *Urban Revitalisation Strategies and Plans in New Orleans 1970–1993*, Working Paper 10, College of Urban and Public Affairs, University of New Orleans.

Lemann, N. (1994) The myth of community development, *New York Times*, 9 January.

Logan, J.R. and Molotch, H.L. (1987) *Urban Fortunes: the Political Economy of Place*, University of California Press, Berkley.

Mahindge (1996)

Marshall Heights Community Development Organisation (MHCDO) (1994) Annual Report and briefing pack: personal interviews with Lloyd P. Smith, Executive Director.

Metrovision (1990) *An Economic Development Agenda for the 21st Century*, Metrovision Foundation Partnership, New Orleans.

Meyer, N.S. (1995) HUD's first 30 years: big steps down a longer road, *Cityscape*, Vol. 1, no. 3, September, Office of Policy Development and Research, HUD.

Modares, A. (1996) Borders to poverty. Empowerment Zones and the politics of development. Urban Affairs Association Annual Meeting presentation, California State University, Los Angeles.

National League of Cities (NLC) (1993) *'All In It Together' Cities, Suburbs and Local Economic Regions*, a Research Report of the National League of Cities, Washington, DC.

National Mortgage News (1996) Homeownership initiative from Clinton, 26 February.

Nelson, R.R. (1996) *Convention Centers as Catalysts for Economic Development: How the Rules are changing in the 1990s*, University of Delaware, Delaware.

Pierce, N.R., Johnson, C.W. and Hall, J.S. (1993) *CitiStates: How Urban America can Prosper in a Competitive World*, Seven Locks Press, Washington, DC.

Schulgasser, D.M. (1995) *Newark's Experience with the Clinton Administration's Urban Programmes*, City of Newark, Newark, NJ.

Stone, C.N. (1989) Regime Politics: *Governing Atlanta 1984–1988*, University Press of Kansas, Kansas.

Swanstrom, T. (1985) *The Crisis of Growth Politics: Cleveland, Kunninoch and the Challenge of Urban Populism*, Temple University Press, Philadelphia.

Urban Institute (1992) *Confronting the Nation's Urban Crisis from Watts (1965), to South Central Los Angeles (1992)*, Urban Institute, Washington, DC.

US Department of Housing and Urban Development (1994) Reinvention Blueprint, 19 December, Washington, DC.

Vaz, K. (1996) *The Future of Urban Policy, City 2020*, House of Commons, London.

The Washington Post (1994) editorial: It's not about makeup, 8 March.

Washington, R.O. (1995) *Revisiting the Concept of 'Citizen Participation' – Implications from the New Orleans Empowerment Zone Planning Process: a Case Study*. College of Urban and Public Affairs, University of New Orleans.

Wiewel, W., Bennington, J. and Geddes, M. (1992) Comparative local economic development: Britain and the United States, Focus issue of the *Economic Development Quarterly*, Vol. 6, no. 4, November.

Wolman, H. and Spitzley, D. (1996) The politics of local economic development, *Economic Quarterly*, Vol. 10, no. 2, May.

Key Contacts

Information on urban matters in the USA can be obtained from:

- Urban Affairs Association
  University of Delaware
  Newark DE 19716
  (302) 831-1681
  E-mail UAA@MVS.udel.edu
- *Economic Development Quarterly*
  Levin College of Urban Affairs
  Cleveland State University
  Cleveland, OH 44115
- National League of Cities
  1301 Pennsylvania Avenue N.W.
  Washington DC 20004
  (202) 626-3000
  The National League of Cities (NLC) serves as an advocate for its members in Washington, represents 1,400 cities directly and provides training, technical assistance and information.
- The Urban Institute
  2100 M Street, N.W.
  Washington DC 20037
- US Department of Housing and Urban Redevelopment
  451 Seventh Street S.W.
  Washington DC 20410

Followers of local economic development and urban policy in the United States need to read the journals *Economic Development Quarterly* or the *Journal of the Urban Affairs Association*. Research institutes like the Urban Institute in Washington have long played a key role in researching the urban crisis and policies, whilst the National League of Cities represents the key players in city governments on Capitol Hill and plays a strategic role in campaigning for a better deal for American cities (see, for example, NLC, 1994).

In Washington also, a number of national community organisations play key roles in community economic development and the Center for Community Change has been unique in providing a resource to community organisations in the last 25 years, working recently with the Anne Casey Foundation to examine sector economic development strategies for job creation (CCC, 1996).

13 European Experiences

Paul Drewe

Introduction

Because there are more than 3,500 towns or cities with a population of more than 10,000 inhabitants in the European Union, it is impossible to provide a comprehensive account of European experience with regard to urban regeneration. As an alternative it is possible to draw on existing comparative studies and, in particular, projects that have received European Union assistance. This can be considered as a sufficiently large and fairly representative sample of local regeneration experiences across Europe. However, the sample may be slightly biased in favour of towns and cities that are either experiencing greater difficulty in coping with urban problems or are adept at obtaining European funding. Nevertheless, the experience of these towns and cities provides a flavour of urban regeneration activities throughout Europe and offers an indication of the benefits of best practice that may be capable of adoption elsewhere.

This chapter focuses on:

- the urban scene at the European level;
- the diversity of responses to urban problems;
- EU activities related to urban regeneration (singling out Urban Pilot Projects and the Urban Initiative);
- the search for good or best urban regeneration practice;
- a plea for an active role for British cities in Europe.

The intention of the chapter is to provide a flavour of the rich and varied experience of urban regeneration on the mainland of Europe and to illustrate this analysis by reference to case studies.

Setting the Scene

About three-quarters of the EU population lives in urban areas and more than half of this urban population lives in cities of more than 200,000. This includes 32 cities with a population of more than a million, among them London and Paris, the only extremely large, or world, cities in the EU. The picture would be incomplete without mentioning the fact that a relatively high percentage of the population live in small and medium-sized towns and cities with 10,000 to 200,000 inhabitants, a fact that tends to be under-estimated by policy-makers.

The trend towards urbanisation is evident in all of the member states of the EU, though it is far from uniform. The north is marked by a slower pace, whereas southern countries are quickly catching up. But as Millan (1994) has pointed out, Europe's territorial organisation is becoming more complex. Concepts such as the centre-periphery dichotomy or the London–Milan axis of growth are no longer adequate to describe an increasingly diversified structure.

European cities in general experience a growing problem of social exclu-sion, aggravated and perpetuated by spatial segregation, in particular the spatial concentration of disadvantaged groups: the unemployed, the young, the unskilled, immigrants and ethnic minorities (European Commission, 1994). Social exclusion is more acute in industrial cities in the north than in cities in the developing south. However, it is important to realise that not only does economic decline cause social exclusion, so does economic growth if certain population groups are unable to share in a rising level of prosperity. Although the causes of social exclusion are manifold, a com-mon underlying factor is a change in economic structures resulting from global competition and technological innovation.

For a detailed analysis of urbanisation, spatial change and the functions of cities in the EU see for example Kunzmann (1993), Parkinson, Bianchini, Dawson, Evans and Harding, (1992) or Presidenza del Consiglio dei Ministri (1996).

Diverse Responses to Urban Problems

Local policy responses to urban problems differ considerably irrespective of whether the problems faced are common to all areas or are specific to a particular area. National urban policy also varies from explicit to implicit approaches (Parkinson *et al.*, 1992) as well as spatial planning systems (European Commission, 1994, pp. 146–158).

Local policy responses take several forms:

- strategic adaptation to urban change in the 'old core of Europe' for example the Hamburg Business Development Corporation, a public–private partnership; the Rotterdam Development Board, a think tank on city–region relations; and the Dortmund experience;
- promoting urban growth in the 'new European core', for example the Montpellier technopole or Barcelona's fourfold strategy, including the creation of municipal companies and institutions to attract and co-ordinate investment; economic development areas, for SMEs; technol-ogy parks in collaboration with a city's universities; municipal/private business creation centres;

- promoting economic growth in the 'periphery of Europe' for example, development of telecommunications infrastructure in Seville or CO-DESPAR (Comité de Développement Economique et Social du Pays de Rennes) in Rennes – an example of consensus building in order to develop a future plan for the city.

More specifically, the problem of social exclusion has been addressed most explicitly in north-west Europe including initiatives in France and The Netherlands (or City Challenge in the UK for that matter). France has adopted the *contrats de ville* approach under which policy the intention is to reintegrate mainly peripheral problem estates into the life of a city through the establishment of large urban projects. The Netherlands has opted for small-scale social renewal in major cities such as Rotterdam as a follow-up to its social housing-led policy of urban renewal (Stouten, 1995). The approach adopted in The Netherlands for the peripheral estate of the Bijlmermeer (Amsterdam) is less top down than the French response (Blair and Hulsbergen, 1993). More recently, the Council of Europe has focused on districts of high concentration of immigrants across Europe (Blair and Hulsbergen, 1995).

A recent comparative study reveals the diverse nature of responses to urban problems (Commission of the European Communities, 1992). This study addresses both national developments and integration schemes within the specific institutional context of the member states of the EU, together with ten specific cases of revitalisation programmes that have been implemented in north-west European cities (Calais, Dortmund, Eindhoven, Belfast, Brussels, Groningen, Charleroi, Mulhouse, Paisley, Bremen).

Studies on the Conservation of European Cities

Another source of European experience is provided by studies on the conservation of European urban areas (Drewe, 1995; European Commission, 1995). These studies demonstrate the diversity of responses to a common problem. The general theme is that there is a need for an integrated approach to conservation which is not simply limited to safeguarding the architectural and historic fabric, but also ensures the integrated revitalisation or regeneration of historic centres or areas. Such a strategy should include opportunities for new economic and social development.

From 16 completed case studies, four basic types of urban regeneration task can be identified:

- revitalisation of run-down historic centres (Charleroi, Cork, Valencia);
- historic centre improvement (Utrecht, Edinburgh);
- revitalisation of old industrial and commercial areas of historic interest (Odense, Athens, Waterford);
- conservation in small and medium towns (Le Puy-en-Velay, Mühlberg, Bautzen, Thiva, Tomar, Viseu, Guadix, Caernarfon).

The study results enable the identification of a number of common themes and innovatory responses, these themes and responses are associated with: approaches, functional integration and project development.

New approaches to urban regeneration include new ways of encouraging active public participation in schemes and projects (such as schemes in Odense, Utrecht and Charleroi) or the use of an impact analysis of tourism and physical environment investment for conservation purposes (Edinburgh).

In achieving integration, Charleroi and Utrecht have developed special policy measures and instruments. In a number of cases links have been established between development and the environment, notably related to the environmental impact of traffic (access and circulation pattern), the natural environment in and around cities (Guadix) and the operation of utility infrastructure networks (Guadix and Tomar).

A variety of conservation projects have been proposed. Historic areas do not necessarily live by tourism alone, and there is considerable scope for innovations in economic approaches and content. Moreover, many new ideas have been developed that are related to the use of soft approaches and infrastructures to support the revitalisation process. There are three outstanding cases of public–private partnership or local synergy: Caernarfon's Business Plan, Cork's Historic Centre Development Trust and the partnership between the pre-existing Edinburgh Old Town Renewal Trust and the Lothian and Edinburgh Enterprise Limited. As far as the phasing of implementation is concerned, some cities have proposed interesting ways of experimenting with, or testing, conservation projects (including simulation exercise, pilot projects and integrated demonstration projects). With regard to impact analysis, Edinburgh's economic impact analysis (even if not yet applied to concrete projects) and Utrecht's sensitivity analysis are worth mentioning. Finally, there are three outstanding examples associated with the promotion and marketing of urban regeneration schemes (Cork, Utrecht, Le Puy-en-Velay).

The relationship between conservation and urban economic development is expressed schematically in Figure 13.1. This relationship suggests that it is possible to identify a model for a conservation project (based upon the outputs of the various studies), which places particular emphasis on investment in buildings and the physical environment (plus their operation and the support of a range of activities). Once implemented, the project is expected to create a primary cash flow related to individual stand-alone functions and a derived cash flow resulting from functional synergy or the combination of different functions (multifunctionality). In addition, the uniqueness of a project, rooted in its historic or symbolic character, can cause 'scarcity' or locational synergy. As is demonstrated by the case of Edinburgh, changes in tourist expenditure in the Old Town multiply, producing direct, indirect and induced effects on total output, as well as income and employment. These impacts can be seen – both in the area and elsewhere in the city. This economic chain reaction can also be triggered by

Figure 13.1 **The economic impact of a conservation project**

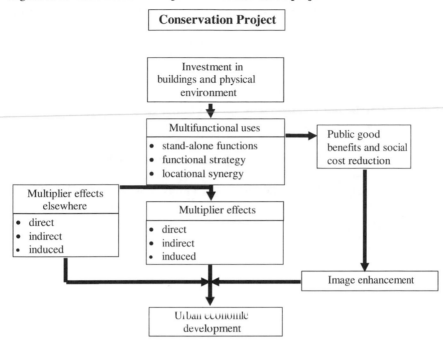

changes in resident, employee and visitor expenditure, depending upon the content of the conservation project.

Above and beyond the economic impacts associated with urban regeneration projects, it is possible to identify social and public-good benefits and social costs. Benefits may relate to historic–cultural values, including effects on non-users related to the symbolic value of a public good, whilst an example of social costs is the environmental damage done by additional traffic that is generated by tourism. The costs of resolving any problems should be internalised through their incorporation in a project budget. Finally, there is another way in which improvements to the physical environment may contribute to urban economic development: this is through the enhancement of the image of an area which can help to promote inward investment.

Urban Regeneration and the European Community

The EU does not have a specific mandate in urban policy and, therefore, it mainly aims to facilitate the exchange of ideas between towns and cities and between member states on how to improve the effectiveness of urban policy. This includes both the transfer of best practice and the development of co-operation networks between cities. Although member states remain responsible for urban policy (this is in line with the principle of subsidiarity) many towns and cities and their institutional representations, such as the Council of European Municipalities and Regions, maintain regular contacts with the various EU services and often seek support for their activities.

As a result, the explicit recognition of an urban dimension in various EU policies is increasing. The Directorate-General of the European Commission primarily in charge of urban matters is DGXVI. It is responsible for regional policy and cohesion. This explains why it focuses on projects which are intended to contribute to the economic functioning and the well-being of urban areas, including the contribution to the development of the region that an urban area is part of. But DGXVI is not the only Commission service involved in urban issues; a broad range of activities is covered by various other services.

Two specific EU activities targeted at urban regeneration can be singled out:

- Urban Pilot Projects;
- the Urban Initiative.

Urban Pilot Projects

During the period 1990 to 1996, 33 urban pilot projects were initiated in 11 member states. A total of 202 million ECU (the ECU is currently worth £0.7) was committed to these projects, with half of this financed by the European Commission (European Commission, 1996). By December 1995, nine projects were completed and an interim progress report has been published (RECITE Office, 1995). A year later, 22 projects had completed their funding periods and had submitted their final activity reports to the Commission. Following the first round, a call for proposals for new urban pilot projects was launched in 1995. It has prompted 503 applications from 14 member states, 26 of which have been selected for support from the Commission in 1996.

At the outset, four main principles were developed to assist in the selection of proposals; these principles required projects to demonstrate:

- a theme of urban planning or regeneration of European interest;
- the innovatory character or new approach proposed by an initiative;
- the demonstration potential of a proposal;
- the contribution to regional development.

In most cases a small part of a city was selected for a pilot project which could lead to solutions to specific problems other than housing or social problems as such.

The 33 projects selected are oriented towards four main themes (although half the projects pertain to more than one theme):

- economic development in areas with social problems: in peripheral and some inner city problem housing areas where lack of access to jobs excludes many from the economic mainstream (Aalborg, Antwerp, Bilbao, Bremen, Brussels, Copenhagen, Dresden, Groningen, Liège, London, Lyon, Marseille, Paisley, Rotterdam);

- environmental action linked to economic goals (Athens, Belfast, Gibraltar, Madrid, Neunkirchen, Pobla de Lillet, Stoke-on-Trent);
- revitalisation of historic centres: bringing back economic and commercial life where for various reasons the inner fabric has been allowed to decay (Berlin, Cork, Dublin, Genoa, Lisbon, Porto, Thessaloniki);
- exploitation of technological assets of cities (Bordeaux, Montpellier, Toulouse, Valladolid, Venice).

What is happening in these projects can best be described in terms of their expected outputs.

- Projects under theme 1 deal with training facilities and services, employment advice, on the job training and the provision of placements, business support services and facilities, new technology products and training, urban planning products, opportunities for socially excluded groups, rehabilitation of buildings, and community services and facilities;
- Projects under theme 2 include: landscape improvements, refurbishment and the reconversion of premises, environmental management products and services, and employment and business opportunities;
- The output of theme 3 projects include: renovation of historic monuments, environmental improvements, reintegration of urban centres into mainstream city life, traffic management improvements, promotion of cultural activities, tourism opportunities and increased local business activity;
- Theme 4 projects relate to: development of technology research, construction of R&D facilities, research institute–industry links, technology training facilities, business support services, reconversion and refurbishment of historic buildings and international networking on research and development.

The 503 proposals for new urban pilot projects cover a larger variety of themes: unplanned suburbanisation, advantages of medium-sized cities, area-based integrated approach, functional obsolescence, social integration, environmental improvements, preserving 'disadvantaged' buildings, access, information technology, institutional constraints and others.

The Urban Initiative

Urban, an EU initiative for urban areas introduced in 1994, focuses on integrated development programmes for the deprived areas of cities. The initiative is part of a European vision of urban areas as part of a more balanced system in terms of economic development, social integration and environment (European Commission, 1994). Cities or urban conurbations with a population of more than 100,000 were eligible for support from this programme. Moreover, priority was given to cities located in Objective 1 or other less-developed regions. The total contribution of the Structural Funds to Urban during the period 1994–99 is 807 million ECU.

The initiative aims to address in an integrated manner the problems of densely populated urban areas which experience high unemployment, a decaying urban structure and infrastructure, poor housing and a lack of social amenities.

Integrated programmes can include:

• launching new economic activities: including the establishment of workshops, support for business, the provision of services to SMEs and the creation of business centres;
• training schemes: language training oriented to the specific needs of minorities, the teaching of computer skills, the creation of mobile units to give advice, and work experience schemes for the long-term unemployed;
• the improvement of social, health and security provision, including: nursery and crèche facilities, drug rehabilitation centres, improved streetlighting and neighbourhood watch schemes;
• the improvement of infrastructure and the environment through: the renovation of buildings to accommodate new social and economic activities, the rehabilitation of public spaces including green areas, the improvement of energy efficiency, and the provision of cultural, leisure and sports amenities.

In order to increase the problem-solving capacity of the various urban areas in question, use can also be made of exchange programmes and partnerships.

By January 1997, 85 projects had been selected (European Commission, 1997). Summary sheets on the operational programmes have been compiled, providing information on the strategy adopted and the results to be obtained, among others. The Brititsh cases cover neighbourhoods in Belfast, Birmingham, Derry, Glasgow, London, Greater Manchester, Merseyside, Nottingham, Paisley, Sheffield and Swansea.

In Search of Good or Best Practice

Urban revitalisation, whatever it is, has a high positive valence; it is seen to be a good thing, worthy of pursuit and emulation. But of what does this good thing consist? Clearer understanding of what people mean by 'urban revitalisation' might also lead to more critical thinking about which aspects of it are indeed 'good' and which are more problematic.

(Wolman, Cook Ford and Hill, 1994, p. 846)

What can we learn from the European experiences? The case studies noted above can provide a source of inspiration for practitioners in Britain and elsewhere. However, it is important to note that the cases in question need to be evaluated, both ongoing and ex post, in their proper context in order to determine whether they qualify as good or even best practice. Even in those cases proven to be successful, it is essential to judge whether the experience can be transplanted or adapted to a different context. Bad practice, on the other hand, is also of help in order to avoid repeating the same mistakes elsewhere.

Good Practice Aspects of Urban Pilot Projects

What works and what does not? An answer to this question can be illustrated by the 32 urban pilot projects for which an interim evaluation has been carried out (RECITE Office, 1995). Not as yet complete evaluations of good practice have been conducted for each theme and some of the key points from these studies are illustrated in Boxes 13.1 to 13.4. The ways in which projects are implemented can be crucial to their success. Several important aspects of the implementation process have come to the fore. Despite the fact that the 32 projects cover a variety of political cultures, a number of common implementation aspects emerge. The four most important features are:

- the extent and quality of horizontal co-operation (local authorities, government departments, local agencies, research institutes, professional bodies and various interest groups);
- the desirability of encouraging vertical co-operation between statutory authorities (central, regional and local);
- the importance of private sector involvement (taking several forms);
- the need to involve local voluntary or residents groups.

In most cases, existing urban regeneration and development agencies, new organisations or specially formed partnerships have been responsible for the design and implementation of projects. In a few cases, project management and co-ordination work has been delegated to independent organisations. In addition, it is important to note that little use has been made of external evaluations by independent consultants.

The interim evaluation is generally a positive one, and the report gives priority to the constructive aspects of the projects. In order to identify the shortcomings of projects, or what does not work, it is necessary to read between the lines. A major problem explicitly mentioned by the evaluators, however, is the fact that 23 of the 32 projects have requested additional time to complete their programme of implementation. Such delays have been caused primarily by the time needed to resolve uncertainties over land ownership, to obtain construction and refurbishment permits, and to establish new organisational structures. More indirect evidence of what does not work satisfactorily can be inferred from the evaluators' suggestions for future pilot projects. These factors include, among others:

- the need for the clearer definition of responsibilities with regard to co-operation between central and local authorities;
- the restrictions imposed by a number of preconditions: no overriding physical or landownership constraints; the absence of effective management and control; the presence of existing detailed proposals; the non-durability of strategic options; the need to build a broad local consensus for projects;
- the need for more clearly defined partnership arrangements;
- a clearer treatment of inputs and financial returns to projects.

The benefits expected from projects should include social returns especially if the lack of employment opportunities is considered to be 'a root cause of urban problems' (RECITE Office, 1995, p. 50). The state of the art of assessing good or best practice in urban regeneration is still under-developed.

A number of lessons can be gleaned from award competitions focusing on, generally speaking, excellence in the urban environment (Langdon, 1990):

- a thorough on-site inspection is required;
- it is not sufficient to report only the good news; shortcomings (which may hold lessons for others) need also to be acknowledged;
- significant assumptions about what constitutes quality need to be made explicit;
- it is essential to examine the artefacts (projects, objects, places – see Box 13.5), and relevant processes and values; process issues include, for example, various aspects of the implementation process mentioned above and an assessment of whether they work or not; values include, for example, intentional diversity (projects serving a broad section of society) or empowerment (enabling people to exert more control over their lives);
- the evaluation should tell the full story of the actors involved – professional, political, social, financial and others – instead of just one type of actor.

Although these lessons are derived from US experience they can be applied to the European situation. For a more academic approach to the evaluation of some European experience see Francesca and Nijkamp (1996).

A Plea for an Active Role for British Cities in Europe

European Union activities in urban matters have a significant and growing influence on British practice, apart from Urban Pilot Projects and the Urban Initiative (European Commission, 1977):

Box 13.1   Economic development in areas with social problems

Good practice aspects (in order of frequency) identified from 14 cases:

- physical impact and improved security;
- training and counselling based on precise targeting of beneficiaries;
- cost-effective subsidies for job placement and on the job training;
- on-site business support facilities;
- actions to improve confidence of persons (pre-training, education, counselling);
- resident-oriented services;
- labour-market oriented training provision and the involvement of employers;
- involvement of beneficiary groups in project design and implementation;
- demonstration effects and transferability of know-how;
- cost-effective subsidised accommodation for small business and crafts.

Box 13.2   Environmental action linked to economic goals

Good practice aspects (in order of frequency) identified from six cases:

- balancing environmental protection with business development;
- demonstration effects and promotion in relation to increased environmental awareness and knock-on effects;
- combining leisure facilities with environmental awareness actions;
- conservation work on premises adapted to business needs;
- impact enhanced with use of 'clean' technologies;
- combining environmental protection with on-site environmental training.

Box 13.3   Revitalisation of historic centres

Good practice aspects (in order of frequency) identified from seven cases:

- high-quality refurbishment standards specified for the restoration of areas and buildings of historic and cultural significance;
- restoration work to adapt a building to new demands;
- traffic improvements to increase public use and improve business opportunities;
- reintegration of historic centres into mainstream city activity involves the clear definition of functional requirements;
- improved environmental standards to increase confidence in locality;
- tourism and cultural opportunities aimed at the attraction of business.

Box 13.4   Exploitation of technological assets of cities

Good practice aspects (in order of frequency) identified from five cases:

- research undertaken corresponds to local industry needs;
- emphasis on high technology transfer to local SMEs and inward investment;
- strong orientation towards commercialisation of research and development output;
- dissemination at local to international levels to enhance business opportunities;
- research in leading-edge technology combined with new qualifications and training;
- physical impact to improve awareness and facilitate dissemination;
- research activity as means to restore functions of historic monuments.

Box 13.5   Urban excellence in products

- Urban buildings are better when they are sensitive to their surroundings.
- Fanciness and originality are not important values *per se*; they can be welcome when they serve a purpose but can be inappropriate or harmful when they do not.
- Preservation of old buildings is one possible component of urban excellence, in part because old buildings enrich a community's sense of history. Preservation is not an absolute value, however; sometimes new buildings are superior.
- Buildings are generally not to be esteemed as objects, but rather as places that make it easier for people to conduct their activities and fulfil their needs.

Many EU policies, in particular the Structural Funds, have a direct or indirect impact on British cities in Objective 1 and 2 regions. Moreover, during the 1990s there have been developments at EU level promoting urban regeneration. These developments include:

- initiatives to support employment (Objectives 3 and 4, EMPLOY-MENT, INTEGRA).
- actions for the urban environment (Green Paper on the Urban Environment, the Fifth Environmental Action Programme, Good Practice Guide to sustainable development in urban areas, European Sustainable Cities and Towns Campaign, LIFE).
- IT applications (Telematics Applications Programme, also focused on issues of urban development: European Digital Cities Project).
- policy studies (such as European 2000+).
- urban networks within the Union (Quartiers en Crise, Eurocities, European Urban Observatory and other networks of exchange of experience and co-operation under the RECITE programme).

- co-operation with Third Countries (Central and Eastern Europe, New Independent States of the former Soviet Union, Mediterranean: ECOS-OUVERTURE, MED-URBS).

The EU is the most important international source of (co-)funding for urban regeneration in Europe unlike other European institutions such as the Council of Europe, Eurocities or the Assembly of European Regions. It can have a pump-priming function enabling activities that would not otherwise have been carried out.

The EU like many other European institutions is also a valuable source of information. However, a central database of best or good practice in urban regeneration does not exist (key information on urban pilot projects is collected at present). Instead of a central database it may be desirable to develop a network of member-state databases, with associations such as BURA providing the national inputs.

It is important to ensure that Britain plays an active role in existing urban networks in the European Union and initiates, if necessary, new networks. The European experience would hardly be complete without the British know-how in urban regeneration and there are also lessons to be learned from diverse responses to urban problems 'on the mainland'. There is also an urgent need for joint ventures, diffusing EU know-how to cities outside the EU, including co-operation with Asian, African and Latin American cities.

References

Blair, T.L. and Hulsbergen, E.D. (1993) Designing renewal on Europe's multi-ethnic urban edge: the case of Bijlmermeer, *Cities*, Vol. 10, no. 4, pp. 283–98.

Blair, T.L. and Hulsbergen, E.D. (1995) Designing and implementing innovative approaches. Background discussion paper prepared for the meeting on area-based projects in district of high concentration of immigrants, Council of Europe, Strasbourg, 16 March.

Commission of the European Communities (1992) Urban social development, *Social Europe*, Supplement 1/92, Luxembourg.

Drewe, P. (1995) *Studies Conservation of European Cities: A Synthesis Report*, Delft University of Technology, Delft.

European Commission (1994) *Europe 2000+: Co-operation for European Territorial Development*, EC, Brussels and Luxembourg.

European Commission (1995) *Studies Conservation of European Cities: Synthesis Report Prepared for the European Parliament*, Directorate-General XVI, Brussels.

European Commission (1996) *Urban Pilot Projects, Annual Report 1996*, EC, Luxembourg.

European Commission (1997) *Europe's Cities, Community Measures in Urban Areas*, EC, Luxembourg.

Francesca, B. and Nijkamp, P. (1996) Cultural heritage and urban revitalisation: a meta-analytic approach to urban sustainability. Paper given at the European Regional Science Association 36th European Congress, ETH Zurich, 26–30 August.

Kunzmann, K. R. (1993) *Defending the National Territory: Spatial Development Policies in Europe in the 1990s*, IRPUD, University of Dortmund, Dortmund.

Key Contacts

Some useful information on EU activities in urban matters can be obtained
from the following sources.

Printed Information

Office for Official Publications of the European Communities
L-2985 Luxembourg

Sales and subscriptions
HMSO Books (Agency section)
HMSO Publications Centre
London

Calls for Tenders

Official Journal of the European Communities
C series 'Notifications and open competitions'

Key Contact Points

European Commission
Directorate-General XVI
Regional Policy and Cohesion
Formulation of regional policies, spatial planning, urban issues, co-ordination
of Article 10 ERDF
Rue de la Loi, B-1049 Brussels
Telephone: +32 2 299 11 11. Fax: +32 2 296 25 68

Internet: http://www.eu.int/comm/dg 16/index_ en.html

RECITE Office (Urban Pilot Projects) managed by
ECOTEC Research and Consulting Ltd
13b Avenue Tervuren, B-1040 Brussels
Telephone: +32 2 732 78 18. Fax: +32 2 732 71 11

Langdon, P. (1990) *Urban Excellence*, Van Nostrand Reinhold, New York.

Millan, B. (1994) Europe 2000+: territorial aspects of European integration, EUREG, *European Journal of Regional Development*, Vol. 1, no. 1, pp. 3–8.

Parkinson, M., Bianchini, F., Dawson, J., Evans, R. and Harding, A. (1992) Urbanisation and the functions of cities in the European Community. Report to Commission of the European Communities, European Institute of Urban Affairs, Liverpool John Moores University.

Presidenza del Consiglio dei Ministri (1996) European spatial planning. Ministerial Meeting on Regional Policy and Spatial Planning, Venice, 3–4 May, Rome.

RECITE Office (1995) Urban Pilot Projects: second interim report on the progress of urban success stories, *Urban Studies*, Vol. 31, pp. 835–50.

Stouten, P. (1995) *Urban Renewal in Transition*, Faculteit der Bouwkunde, Technische Universiteit Delft, Delft.

Wolman, H.L., Cook Ford III, C. and Hill, E. (1994) Evaluating the success of urban success stories, *Urban Studies*, Vol. 31, no. 6, pp. 835–50.

14 Current Challenges and Future Prospects

Peter Roberts and Hugh Sykes

Key Issues

Previous chapters of this book have provided a variety of insights into the evolution and current state of urban regeneration in the UK, mainland Europe and North America. This chapter provides a synthesis of what has come before and, building on this, offers a view of the possible future evolution of urban regeneration.

In offering this summary and glimpse of the future, the authors are aware of the dangers that are inherent in attempting either to distil the vast range of present-day experience into a single summary of current challenges, or develop a definitive view of future prospects. Indeed, these concerns are of such significance that the best that the present authors can claim is that what follows represents their opinion of what currently exists and what might develop in the future.

Despite these self-imposed limitations, this chapter attempts to provide a comprehensive overview and judgement of the present state of urban regeneration theory and practice. It also considers a range of views regarding the challenges of the future. In particular, the chapter examines three issues:

- the major distinguishing features and characteristics of current practice;
- the ways in which urban regeneration adds value, the distinctive contribution made by urban regeneration, and the strengths and weaknesses evident in present-day urban regeneration theory and practice;
- the future evolution of the 'urban challenge' and the response of regeneration policy at both urban and regional levels.

The Features and Characteristics of Current Practice

In this section attention is focused on the key distinguishing characteristics and features of urban regeneration. Whilst some would claim that urban regeneration is simply any *ad hoc* proposal, action or inducement that brings about change in the circumstances of an urban area, this is not the definition that has been adopted in this book and it does not reflect the inherent qualities evident in the best of urban regeneration practice. As was stated in Chapter 2 of the present text, urban regeneration can be defined as:

> comprehensive and integrated vision and action that leads to the resolution of urban problems and which seeks to bring about a lasting improvement in the economic, physical, social and environmental condition of an area that has been, or is, subject to change.

Whilst this definition casts the net wide, it does help to establish criteria by which to judge any plan or action that claims to be involved in urban regeneration. For example, 'patch and mend' policies that tinker with the provision of social infrastructure, and which take place in isolation from the mainstream of economic, environmental and social policies in an area, can hardly be described as urban regeneration.

This suggests that three issues should be considered in any attempt to evaluate the adequacy of an individual urban regeneration initiative:

- the nature of the challenge encompassed by the term 'urban regeneration';
- the approach adopted in order to meet the challenge;
- the outcomes of the application of the approach adopted.

A useful summary of some of the most important characteristics of urban regeneration has been provided by Parkinson (1996), who argues that, among other factors, the search for successful urban regeneration has been stimulated by the need to address problems associated with:

- a rapidly changing economic environment in which there has been an increase in the range of problems facing many cities, but a reduction in the level of public or private sector control that can be exercised over economic decisions;
- the loss of well-paid manufacturing jobs, a growing division between well-paid and poorly-paid jobs in the service sector labour force; an increase in part-time employment and a shift in the gender mix of jobs;
- new social trends resulting from demographic change; the breakdown of traditional family structures; the decentralisation of people and jobs; the move out of the city of the younger and more able population; the loss of social cohesion and the transformation of traditional communities; the creation of new communities, including the roles played by ethnic groups.

A number of other problems can be added to this list, including the continued deterioration of the physical state of many urban environments; the need to take urgent action in order to avoid permanent harm to human health and the need for costly environmental rectification measures in the future; the physical decay of towns and cities that results in a serious underutilisation of scarce resources and creates pressure for the expansion of urban areas; and the decay or obsolescence of much urban social and economic infrastructure.

The following paragraphs of this section examine three issues:

- the features that distinguish urban regeneration from other associated activities;
- the characteristics that define best practice;
- the contribution made by urban regeneration to associated areas of interest and activity.

Distinguishing Features

Urban regeneration can be distinguished from other forms of urban intervention and policy by reference to a number of features. It is:

- essentially a strategic activity;
- focused around developing and achieving a clear vision of what action should take place;
- concerned with the totality of the urban scene;
- engaged in the search for both short-term solutions to immediate difficulties and long-term approaches that anticipate and avoid potential problems;
- interventionist in approach, but not *dirigiste* by nature;
- best achieved through a partnership approach;
- concerned with setting priorities and allowing for their achievement;
- intended to benefit a range of organisations, agencies and communities;
- supported by various sources of skill and finance;
- capable of being measured, evaluated and reviewed;
- related to the specific needs and opportunities present in an individual region, city, town or neighbourhood;
- linked to other appropriate policy areas and programmes.

Although when considered collectively the above-noted features can be seen to be associated with the practice of urban regeneration, when taken individually they are also representative of many other types of activity. What is unique to urban regeneration is the combination of these features and their application through an integrated package of measures to the resolution of problems in the urban domain. However, this holistic approach does not suggest that urban regeneration is a fixed discipline. It is not, and reminding ourselves of the lessons of the past, it would be unwise to consider an individual urban problem as an issue that can be addressed solely within a particular urban area. As Hall (1981) argued many years ago, most individual urban problems should be considered in the context of the metropolitan area, region or nation in which they occur, and it is clear that the nature and impact of such problems can change considerably over time.

This discussion leads to the consideration of what is the most appropriate scale of action for urban regeneration, and here there is little chance of providing a straightforward answer. Over the last thirty years of the twentieth century the scale considered appropriate for the definition of urban problems and the design of urban policy has shifted from the very local to the city-region, to the neighbourhood or district, and back again to the region.

Whilst some governments have favoured a micro-area approach to urban regeneration, others have applied more broadly based policies. What is right and wrong is not the issue here, rather it is important to acknowledge that urban problems vary in terms of their cause, character and occurrence. A suitable policy framework is one in which each problem can be addressed at a suitable spatial scale. Some problems, such as the provision of links to the international transport system, cannot be tackled effectively at a very local level, whilst neighbourhood concerns are best dealt with locally. It is clear, however, that local initiatives alone are unlikely to be sufficient to overcome major structural difficulties (Pacione, 1997), whilst broad-brush national solutions may lack the fine cutting edge that is vital for successful policy design and implementation at regional or local level. The selection of a suitable spatial scale for the development of an urban regeneration programme or project is akin to unpacking a Russian doll: each level of policy must be considered and appropriate acknowledgement should be given to the other layers of policy both 'above' and 'below' the specific activity which is the focus of concern.

The Characteristics of Best Practice

An important question to consider in any investigation of the performance of urban regeneration is: how can best practice be identified and what lessons can be obtained from the study of the key characteristics of such practice? One problem here is that there is no single or precise agreement as to what constitutes best practice and, as a consequence, what is considered to be excellent in one locality or sector of activity may be disregarded elsewhere. Lawless has argued that this problem reflects a lack of primary research and literature, especially related to the practice of urban policy and the lessons that can be gleaned from study of international experience (Lawless, 1995), whilst other observers have noted the difficulty of drawing early definitive conclusions on the success of what are essentially long-term efforts to encourage regeneration (Geddes and Martin, 1996).

Although urban regeneration as an art and a science is still in its infancy, and accepting that any definitive assessment of best practice in urban regeneration will have to be delayed until an independent assessment has been conducted, some interim lessons can be obtained from a study of the British Urban Regeneration Association's Best Practice Awards. This award for best practice in urban regeneration has been offered for the past seven years, and the award winning schemes represent a range of different approaches to urban regeneration, drawn from various areas of the UK. The criteria used to judge nominations for the BURA Best Practice Award echo the call made by Oatley (1995) for urban regeneration to adopt a comprehensive and integrated view of urban problems and to bring forward solutions that offer a long-term strategic approach. In judging a scheme or project that has been nominated for a Best Practice Award, the assessors consider:

- the contribution made to the economic regeneration of an area and the financial viability of the initiative;
- the extent to which a scheme has acted as a catalyst for further regeneration in an area;
- the contribution made to community spirit and social cohesion;
- the contribution made to building the capacity of local people to plan and influence the future development of their area;
- the environmental sustainability of a scheme or project;
- evidence that points to the success of a scheme in the past, at present and into the future;
- the range of partners involved in a scheme;
- the presence of a concern for the longer-term development and management of a scheme;
- qualities of imagination, innovation, inspiration and determination.

The Wider Contributions of Urban Regeneration

As has been suggested above, urban regeneration does not stand alone. Most schemes and projects either are part of a wider programme of action concerned with the overall improvement of an urban region, or make a contribution to adjacent spheres of activity. The wider contribution of urban regeneration can also be seen as an important element in the process of national and regional development and regeneration.

Given the wider contribution made by urban regeneration, it is essential to consider individual schemes in the context of the broader socio-economic and physical environments within which they operate. This evaluation of the broader environment and the wider potential contribution of urban regeneration provides a basis for calculating the value which is added by urban regeneration. At the heart of any such assessment is an estimate of the costs and benefits associated with an initiative and the assignment of these costs and benefits to the various parties and partners who have been involved. This approach to assessing the contribution of urban regeneration is reflected both in practice (for example, many major international exhibition and tourism development schemes make a contribution to regional and national facilities as well as providing for local needs) and in the various evaluation studies that have attempted to trace the impact of urban regeneration in general, and of major projects in particular (Loftman and Nevin, 1995).

Added Value and Strengths and Weaknesses

Although urban regeneration as a general approach is now an established facet of the urban and regional policy scene, it is also evident that the nature and content of regeneration practice has changed considerably during the 1980s and 1990s. This constant process of transformation in the nature and content of regeneration is nothing new. As Chapter 2 of this book argued, urban regeneration, like many other aspects of spatial and sectoral policy, is heavily influenced by the events and circumstances which confront it. As a consequence, the nature and content of urban regeneration has had to change in order for it to remain relevant and effective.

This suggests that it is important to be able to define and distinguish those areas of urban regeneration practice that are of particular importance for local, regional and national development. The overall purpose of this exercise is to isolate those aspects of urban regeneration that may provide a model for future policy development and implementation. Three issues are considered in this section:

- the role of urban regeneration in policy development and implementation;
- the value added by urban regeneration;
- the strengths and weaknesses of current theory and practice.

The Role of Urban Regeneration

Urban regeneration comes in many guises, performs many roles and can help to bring about a wide variety of changes. The extensive array of roles and purposes of urban regeneration reflects the wide range of issues that fall within its scope. In particular, urban regeneration aims to address the various forces and factors that bring about urban degeneration and to prepare a positive and lasting response that results in a permanent improvement in the quality of urban life. Urban regeneration can be seen to perform a variety of tasks including:

- the provision of a framework for the analysis of urban problems and the identification of development potential;
- the generation of an overall strategy for an area and the provision of detailed schedules of implementation and action;
- the identification of the constraints, opportunities and resource requirements associated with a regeneration proposal;
- the establishment of a framework for action, including management arrangements, the assignment of responsibilities and the identification of resource inputs;
- the negotiation of a 'contract' for the establishment and operation of a partnership that has responsibility for the above issues;
- the monitoring, review and roll forward of all the above roles.

Partnership, strategy and sustainability form a troika of approaches that determine and drive successful regeneration. They enable urban regeneration to be more than the sum of its constituent parts and they provide a basis for comprehensive and integrated action. These roles and characteristics have figured significantly in this book and they represent the foundation stones upon which the wider contribution of urban regeneration has been built. Each of these factors can be seen to represent a specific role and to make a particular contribution.

Partnership
Urban regeneration has provided a laboratory for the development of partnership. The move from state provision to private sector-led urban regeneration which took place during the early 1980s initially caused a considerable degree of conflict, confusion and concern, especially with regard to the vexed question of local accountability. At the same time, central–local government relations were subject to a series of adjustments, often with the result that the centre increased its control over spending and policy (Oatley, 1995). From these new policy arrangements, with all their imperfections and tensions, the partnership model emerged (Chapman, 1998). Other factors which have influenced or encouraged the emergence of the partnership model in the UK, include the transmitted experience of similar organisational structures from the United States and elsewhere in Europe, the gradual development of grass roots partnerships and the contributions made by voluntary sector organisations, and the general realisation that partnership offers a valid response to a shortage of resources or a lack of power.

Some of the key lessons of partnership in the urban regeneration field that are relevant and applicable to other spheres in urban and regional policy include:

- the importance of establishing and maintaining an open and equal partnership;
- the need to provide effective and accountable leadership;
- the desirability of creating partnerships that have a long-term strategic purpose rather than simply providing an impermanent alliance that has been hastily constructed for the purpose of securing funds;
- the need to develop an overall strategy that can guide and direct partnership efforts, be used to determine the resource contributions to be made by the partners and to assign responsibilities to individual partners or outside agencies;
- the maintenance of open and accessible records and the provision of regular briefings on progress made and future intentions;
- the desirability of networking with other partnerships and other organisations and authorities – it is all too easy to assume that a partnership encompasses all of the essential views and participants;
- the need to manage and direct the dynamics of a partnership – with the best will in the world enthusiasm will not translate itself into action;

- the need to think about exit strategy or continuity arrangements;
- the desirability of distilling and disseminating good and best practice from the operation of a partnership.

Strategy

Strategy is the second distinctive role and contribution of urban regeneration in the wider urban policy arena. Unlike earlier *ad hoc* attempts to develop and implement urban policy, urban regeneration is an activity that seeks to provide an all-embracing, lasting and comprehensive solution to urban problems. This implies that urban regeneration provides a strategic approach rather than simply offering a collection of unconnected interventions and actions.

As was suggested in Chapter 3 of this book, urban regeneration provides a 'spine' for urban policy intervention. This strategic role has also produced a range of other benefits, including the provision of a framework that can be used to guide associated programmes of action, the establishment of a basis for the further definition of the roles and commitments developed within a partnership, the introduction of an approach that can help in the planning of individual projects, and the provision of a method that can help to ensure the efficient and effective use of resources. This model of strategic urban regeneration offers considerable capability and potential that can be applied in other policy fields such as regional development and rural regeneration. Strategy is fundamental to urban regeneration, and clear strategic vision is likely to continue to be a hallmark of successful regeneration schemes. However, strategic vision also implies the need for strategic resource commitments and this essential element for successful urban policy frequently still appears to be beyond the delivery capabilities of many public and private sector bodies – the vision can be provided but the commitment is often absent.

Thinking and acting strategically requires both confidence in the intended purpose and outcomes of a regeneration scheme and the ability to command and direct resources. These qualities reflect many of the most important characteristics of urban regeneration and demonstrate the need for urban regeneration to operate across the boundaries that frequently divide economic objectives from environmental concerns and social issues. This suggests that strategy is also a major element in providing the basis for the third role and contribution of urban regeneration: sustainable development.

Sustainability

The third role and contribution of urban regeneration is the development and application of an approach to the resolution of urban problems that places particular emphasis on the need for solutions to be sustainable. This book has adopted the standard definition of sustainable development which suggests that urban regeneration should promote the balanced development and management of the economy, society and the environment. Sustainable development also places particular emphasis on safeguarding the interests of future generations and upon the equitable distribution of costs and benefits. In addition, sustainable urban regeneration should place particular emphasis on the promotion of new economic activities and jobs that enhance environmental quality. This can be achieved through the adoption of an ecological modernisation approach (Roberts, 1997a).

In order to ensure that urban regeneration is sustainable, it is essential that all of the component activities should be subject to rigorous screening and evaluation. However, given its partnership base and strategic approach, it is equally evident that urban regeneration can perform an enabling role in ensuring the implementation of sustainability. A number of the policy priorities that the OECD (1990) has defined in relation to the promotion of sustainable development in urban areas are fully coincident with the priorities of urban regeneration. Furthermore, in many areas urban regeneration has become either a champion of sustainable development or an important means for the generation and implementation of sustainability policies. These are important matters because, as the Brundtland Report argued: 'The future will be predominately urban, and the most immediate environmental concerns of most people will be urban ones' (World Commission on Environment and Development, 1987, p. 255).

The Value Added by Urban Regeneration

Above and beyond the individual roles and contributions that are associated with urban regeneration, it is also important to identify the overall value which is added through the adoption of an urban regeneration approach to the resolution of urban problems.

As was demonstrated in Chapter 10, a number of attempts have been made to identify and measure those elements of urban regeneration policy that make a particular contribution to the quantity or quality of urban activity. However, many of the techniques and methods that have been employed to evaluate urban regeneration policy exhibit a tendency to emphasise the contribution of directly measurable outputs. This has resulted in two weaknesses becoming increasingly evident in many evaluation exercises: a tendency to confuse effectiveness with efficiency, and an absence of any real attempt to gauge the overall consequences or results that flow from regeneration efforts. As in judging ice skating, yes it is important to assess 'technical merit', but it is equally important to be able to evaluate and judge the overall lasting 'artistic impression', or value for money, of urban regeneration. The latter quality is likely to represent the true and lasting value of regeneration effort. The real role and contribution of urban regeneration is far more than can be expressed simply in terms of the input of finance or the output of Treasury-approved deliverables.

A number of observers have provided insights into the overall value which is added by urban regeneration. Brian Robson, for example, has argued that 'the "easy" tasks of improving the environment – of building new buildings and refurbishing the old, of installing new infrastructure or cleaning up the physical environment – have been better met than have the

more difficult tasks of creating jobs and strengthening local economies'
(Robson, 1995, p. 48), whilst other wider successes – engaging the local
community and tackling social, environmental and economic problems –
have been reported by others (Beecham, 1993). Studies that offer evidence
of the successful and lasting comprehensive treatment of the problems
encountered across a large area of a British town or city are much rarer
beasts. Little victories have been recorded – the Eldonians Village in Liverpool offers one example, whilst another is the regeneration of previously
unloved peripheral estates such as Whitfield in Dundee – but there are, as
yet, no fully evaluated longitudinal studies of integrated action across an
entire conurbation. Research in other countries, such as that reported in
Chapters 12 and 13, provide a number of clues that can be used to guide
further work in Britain, but at present the precise calculation of the long-
term overall added value of urban regeneration remains a matter of
speculation.

So how will we be able to recognise successful urban regeneration when
it emerges and will the effort involved be matched by the results? The
answer to this question can be constructed in several ways: first, as a statement of aims and aspirations, second, as a projection forward of current
achievements or, third, as a view of the likely future condition of urban
areas and urban policy. The first of these issues has already been considered in this and previous chapters, whilst the third issue forms the basis
for the final section of this chapter. The second issue – the projection
forward of current achievements – can be seen as a reflection of the
strengths and weaknesses evident in present urban regeneration practice.

Strengths and Weaknesses of Current Practice

As noted above, an assessment of the strengths and weaknesses of current
practice can help to provide a foundation for the future progress of urban
regeneration. It can also help in determining if it has provided a lasting
solution to the problems associated with urban degeneration.

Taken overall, and having regard to the many individual positive and
negative variations from the average situation, the strengths and weaknesses of current urban regeneration theory and practice can be identified
both from the general literature and from the personal experiences of the
authors who have contributed to this book. The following sections offer a
brief summary of some of the main elements of both topics.

Strengths

Based on the assessments provided by authors such as Oatley (1995), Shaw
and Robinson (1998), Parkinson (1996) and Burton (1997), and on the
experience of the present authors, the strengths of the urban regeneration
approach encapsulate many of the features of best practice discussed
above, including:

- the provision of a comprehensive and robust long-term integrated strategy for the regeneration of a neighbourhood, quarter, district, town, city or metropolitan region;
- the incorporation of the economic, social, environmental and physical aspects of regeneration in an overarching strategy and detailed programmes of action;
- the development and implementation of a strategy and programme through a partnership approach that involves organisations and individuals from both within and outwith an individual area;
- the agreement of a basis for the provision of the necessary leadership, management and participation arrangements that are essential for a partnership to be effective;
- the definition of priorities and targets within an overall framework that also sets a timetable, provides budgets and assigns responsibilities;
- the provision of an agreed means of monitoring, reviewing, evaluating and revising the strategy and programme of action in order to take account of the changing internal situation and the evolution of external circumstances;
- the specification, either at the outset or during the course of a regeneration programme, of either exit arrangements (where appropriate) or an agreement on future local/community ownership and control;
- the identification, evaluation and dissemination of best practice from a scheme or project.

Table 14.1 **BURA Best Practice Awards 1992 – 1998**

County or Region	Community initiative	Quarter or neighbour-hood	Reclamation or refurbish-ment	Town centre programme	Commercial and industrial	Total
N. Ireland	1	—	—	1	—	2
Scotland	2	2	1	1	—	6
Wales	—	1	1	—	—	2
North	—	—	1	—	—	1
North West	2	2	2	—	1	7
Yorkshire	1	2	2	1	1	7
West Midlands	1	1	2	2	2	8
East Midlands	1	—	—	—	—	1
South West	1	—	—	—	1	2
South East	1	1	1	—	—	3
East Anglia	—	—	—	—	—	0
TOTAL	10	9	10	5	5	39

Even though this summary of the strengths of urban regeneration presents a generalised picture, it reflects many of the characteristics and features of practice that have been identified in the schemes nominated for the BURA Best Practice Award. These schemes can be seen as representative of the wider base of urban regeneration practice which is in evidence in many areas of the UK. In addition, the schemes also reflect the many different forms of urban regeneration action that exist and the various organisational mechanisms that are used to promote and progress regeneration programmes. A summary of the locations and types of urban regeneration scheme that have been successful in gaining BURA Best Practice Awards is provided in Table 14.1.

Weaknesses
Many of the features and qualities noted above can be reversed and considered as weaknesses – for example, the absence of a strategy or genuine partnership arrangements is likely to undermine and damage most urban regeneration efforts – or as matters that require attention in order to prevent them from becoming weaknesses. In addition, a number of other weakness or flaws in the theory and practice of urban regeneration can be identified, including:

- the absence of an adequate or complete definition, understanding and policy position with regard to the origin, occurrence and likely outcomes of an 'urban problem';
- the lack of a clear or consistent position regarding the role, structure and operation of regeneration policy at national, regional, metropolitan or local level;
- the imposition of unrealistic or inflexible planning and other policies that may restrict the potential development of an area;
- the fragmentation of responsibilities and a lack of co-ordination in the design and discharge of policy and implementation;
- an overemphasis or overreliance on a single activity, sector or policy instrument;
- the unnecessary exclusion of a key group or organisation from a partnership;
- the problem of bureaucratisation and the danger of requiring/adopting overelaborate and complex management and organisational structures;
- the lack of a strategy or a commitment to long-term action;
- the absence of an open, transparent and accurate means for recording and evaluating the outcomes associated with a scheme or project;
- the operation of a scheme or project in isolation from other aspects and examples of urban regeneration – this will prevent the input of prior experience from elsewhere and will limit the contribution of new best practice to the overall store of knowledge.

Although the presence of an individual weakness may not prove to be fatal to an urban regeneration scheme, the cumulative effects of a

concentration of these weaknesses can undermine the process of regeneration. This is not to suggest that small or limited projects that are short-term,
topic-specific and within the competence of an individual organisation
should not proceed, rather the key message is that most urban regeneration
schemes and projects are likely to be more successful if they can avoid the
weaknesses that have been identified above. A particular requirement is
that regeneration efforts should be comprehensive, integrated and directed
by strategy (Carley and Kirk, 1998).

What is most important in considering the future of urban regeneration
is that the weaknesses of current practice should be addressed and that the
lessons of best practice should be recognised and accepted. Building upon
the basis of best practice will hopefully help to resolve difficulties before
they become problems.

Such expertise can also help to identify and provide concrete examples
of the application of urban regeneration theory and practice that are of
great value in the education and training of urban regeneration specialists.
The quality of urban regeneration management is variable and the need for
specialist training and retraining programmes is clear. Universities have an
important role to play in supporting this aspect of urban regeneration
through the provision both of initial training and of continuing professional
development. In addition, it may be worth while to move towards the
establishment of a common professional qualification in local and regional
regeneration.

The Future of Urban Regeneration

In considering the future of urban regeneration in the UK, it is essential
both to take account of the likely evolution of the 'urban problem' and to
anticipate the possible future development of policy instruments and structures. Whilst the former issue represents the challenges to which urban
regeneration will have to respond, the latter point reflects the priorities and
field of action across which regeneration policies will operate.

This final section considers two major issues:

- the future challenges and choices that will confront urban regeneration;
- the likely future evolution of urban regeneration policies, structures
  and approaches.

Future Challenges and Choices

As was noted in Chapter 2, towns and cities are the subject of constant
change. Even the most remote or 'protected' urban region is not immune
from the forces that bring about change, and the outcome of an initial
round of adjustment usually acts as the trigger for further change. There is

no reason to suppose that the future will differ from the past in this respect. Indeed, it is likely that the pace of change will quicken, especially as a consequence of further rounds of technical innovation and the search for new styles and modes of urban management, governance and living (Brotchie *et al.*, 1995).

Furthermore, it is important to appreciate that the problems confronting British towns and cities, and those elsewhere in Europe, are relatively insignificant when compared with the current and likely future state of cities in many parts of the Third World. However, cities in both the developed and developing world are changing rapidly and, according to Michael Cohen (1996), they are converging at least in terms of the problems which confront them and the policy instruments that are deployed in response to such problems. It is argued that cities are becoming more alike, despite the continual search for competitive advantage and new ways of responding to the challenge of change, and this growing similarity between cities boosts the need for the widest possible exchange of knowledge and experience.

However, despite this tendency towards convergence at the macro-level, towns and cities also continue to display considerable internal diversity. Indeed, it has been argued that in recent years economic change and social polarisation have extended internal diversity, including the creation of extensive social exclusion which threatens 'the legitimacy of the political system' (Jewson and MacGregor, 1997, p. 9). As a consequence, it is both unrealistic and undesirable to consider the urban future as a single uniform scenario.

Within the urban regions of Britain it is likely that these processes of change will bring about a number of adjustments in the rank order of 'good places' and their constituent neighbourhoods. Once prosperous, pleasant or popular places can slip into degeneration, whilst other towns and cities may experience regeneration and revival. Such changes can have as much to do with image and promotion as with physical regeneration (Shaw and Robinson, 1998), whilst the stimulation of regeneration through the introduction of new sectors and activities that were previously underrepresented in a city can help to bring about the required transformation (Landry *et al.*, 1996).

The current diversity between cities that distinguishes 'good' from 'bad' places is also reflected within cities. Distinctions between neighbourhoods and quarters are as significant as the differences between towns and cities; the most acute manifestations of these internal differences can be seen in the social polarisation evident in, especially, certain peripheral estates and inner city neighbourhoods. Despite the move at macro-level towards a greater degree of convergence, within the city divergence still dominates and the urban spatial mosaic becomes more complex. The real danger is the emergence of the 'doughnut city' and, in order to avoid this, every encouragement should be given to the reuse of brownfield land within urban areas.

There have been few attempts made in the UK to project the present situation forward at a level of disaggregation that allows for detailed strategy to be established. Although such attempts as do exist are somewhat dated, they demonstrate the importance of considering the future of a town or city through the construction of small-scale district or neighbourhood scenarios (see, for example and greater detail, Thew, Holliday and Roberts, 1982; Schnaars, 1987). Typically, such studies consider alternative pathways of economic growth, socio-political attitudes, physical change and intervention, external policy constraints, etc., and from these exercises they construct alternative scenarios that can then be applied to the constituent areas of a town or city. These multi-pathway scenario models are far more helpful than unidirectional or single condition models that classify and depict entire urban systems as a single entity. There is little that is novel in such models, all that they really offer is a projection forward of the present realities of neighbourhood differences and spatial segmentation.

So what challenges will confront urban regeneration in the future? Three issues are likely to dominate the agenda:

- the need to tackle questions of economic development and social justice through the design and implementation of a comprehensive approach that maximises and secures economic progress and reduces the incidence of social exclusion – this emphasises the need to work with and alongside communities to determine the future, rather than imposing solutions from outside;
- the need to ensure that a long-term and integrated strategic perspective is established in relation to the development of urban regeneration policy and the introduction of procedures and processes for the implementation of strategy;
- the adoption of the goals and aspirations of sustainable development, in general, and of environmental sustainability, in particular.

In addition to the above issues, there are a number of other aspects of policy that will be rolled forward, albeit in new guises or with new emphases; they include:

- the desirability of providing a more satisfactory spatial and social context for urban regeneration in order to provide a more complete basis for the generation of individual schemes and projects, and in order to ensure that the benefits of urban regeneration are distributed to the intended recipients;
- the refinement of methods and procedures for brokering and managing partnerships and for ensuring the involvement of local communities;
- the introduction of improved procedures for determining resource requirements, availability and shortfalls, and for ensuring that any shortfalls are addressed and resolved prior to the commencement of an urban regeneration scheme or project;

- the improvement of existing methods and techniques for physical regeneration, the provision of transport, utility infrastructure and other
'hard' elements of regeneration;
- the provision of enhanced methods of monitoring, review, evaluation
and accountability;
- the establishment of a more precise mandate for any organisation or
agency involved in urban regeneration – this should be set within the
context of enhanced urban governance.

The challenges that confront urban regeneration will, of course, vary
from place to place and over time, and in different places, different priorities will be agreed and implemented. What works in one town or city
will reflect such choices and the opportunities that are available. However,
there are a number of common problems that are likely to appear in most
localities, and these common elements will remain as the core issues that
urban regeneration will have to address. Providing jobs, homes and quality
of life in safe and environmentally sound urban areas are universal tasks,
and they represent the constant elements that are at the core of regeneration. Above all else, towns and cities must confront the real issues and
make difficult choices – it may be necessary, for example, to facilitate car
access in the first instance in order to help to revive a declining commercial
centre, and then to move to a more environmentally sound form of transport when resources permit. What is needed is a long-term vision within
which specific pathways can be identified and followed.

Policies, Structures and Approaches

As has been discussed in many places elsewhere in this book, urban regeneration has evolved from old-style urban reconstruction and renewal – a
process that tended to follow a standard pattern of extensive clearance,
rehousing (often on peripheral estates) and town centre development – to
the practice of the present day. Even when compared with the average
practice of the mid-1980s, current urban regeneration reflects the incorporation of new ideas that have resulted in the refocusing of effort on a wider
range of social and environmental concerns and the greater direction of
attention to the provision of long-term strategic solutions. The final subsec-
tion of this book considers the future form and structure of urban regeneration and speculates on what urban regeneration practice may look like in
2010.

Three characteristics can be identified that will be of particular importance in the future practice of urban regeneration:

- the three key issues referred to in the previous section – the need for a
comprehensive approach that deals with economic and social issues, the
provision of a long-term integrated strategic perspective and the adop-

tion of the goals of sustainable development – will define the nature, content and form of urban regeneration theory and practice;

• the field of action within which urban regeneration operates will be determined at a regional or sub-regional level – this will allow urban regeneration to better manage many issues such as the transmission of benefits to the intended recipients, the establishment of a 'balanced portfolio' approach to regeneration and the integrated treatment of urban and non-urban issues;

• partnership will continue to be refined both as a concept and as a means of extended urban governance: particular emphasis will be placed on the development of institutional mechanisms for the incorporation of community-based inputs, for the introduction of greater accountability, and for the continuation of joint funding through financial mechanisms such as PFI and community-based schemes such as credit unions and LETs.

Taken together, these issues represent a new agenda for urban regeneration based upon the major lessons from the past. Some of the most important lessons have been summarised by Shaw and Robinson (1998) in the following terms:

• physical transformation is only part of the regeneration process;
• everything is interrelated;
• the trickle-down effect does not always work;
• regeneration is too important to be left to non-elected quangos;
• partnerships are vital but need to be sustainable;
• resources are never sufficient;
• it is important to have clear aims and realistic objectives;
• image matters;
• regenerating people, rather than places, is difficult to achieve;
• sustainability is the key.

Much of what has been presented in this chapter encapsulates the key lessons from past experience and this will help to inform future practice. In addition, before the 'Postscript' and 'Finale', two of the three aspects of future practice that were presented above require further discussion: the question of spatial scale and the role of partnership.

Spatial Scale: from Urban to Regional Regeneration

Urban regeneration in the past has generally been somewhat restricted in terms of its spatial mandate and field of action. A criticism made of many of the initiatives that have been introduced during the last three decades of the twentieth century is that they have covered too small an area and too restricted a list of subjects to have a real impact on the problems of urban degeneration. In addition, it has been argued that having a 'national' urban policy for England has created a situation in which a standard, nationally determined, solution has been imposed on localities and regions, irrespective of whether this standard approach meets the needs of individual areas or represents the best use of resources. In future, the establishment of an integrated strategic initiative for regeneration at regional level would appear to offer the prospect of better relating individual urban regeneration schemes to a wider regional context, and of providing a basis for bringing together all the necessary aspects of policy and implementation in a single portfolio. An integrated regional strategic approach will also allow urban regeneration to make a more effective contribution to both overall regional regeneration and the comprehensive treatment of individual problem areas (Roberts, 1997b).

The Regional Development Agencies, together with a number of other new organisational structures at both regional level and within local communities, offer the promise of extending what Parkinson (1996, p. 17) has described as 'a more strategic approach in Scotland' to the English regions. However, in addition to these new inter- and intra-regional structures and initiatives in England, it is also clear that urban policy and regeneration is becoming more important as an item on the agenda of the European Union. Action at European and UK levels to reinforce and support a wider context and framework for urban regeneration in the regions of England, Wales, Scotland and Northern Ireland offers the prospect of more consistent financial planning in the future and the promotion of greater diversity in terms of regeneration policy and instruments. These new potentials provide a basis for the provision of more appropriate policy and action that is better tailored to the needs of individual towns and regions. This opportunity will be more likely to succeed if it can be directed through an overarching single regional programme (Roberts, 1998).

Furthermore, and reflecting the wider debate on the desirability of meeting the predicted land requirement that is considered necessary in order to accommodate future residential and non-residential growth, the establishment of greater regional autonomy and competence will enable urban regeneration schemes to either encompass, or be directly linked to, the creation of 'balanced portfolios' of land at regional and local levels. Such portfolios will include both previously used and greenfield sites, and will be aimed at meeting the likely demand for sites for residential, industrial, commercial and other uses. The key to the development and implementation of a 'balanced portfolio' will be a carefully designed regeneration strategy.

The above observations reflect the view that the resolution of urban problems is a matter of concern for everyone. It is not a matter of choice, an individual cannot opt to be isolated from urban problems, the only real decision to be made is how best to address the problems. By setting urban regeneration policy and action within a regional context, greater benefits can be derived, both for the intended recipients of an individual regeneration scheme and for the region as a whole.

Partnership as a Permanent Feature

Finally, there is the question of the future development of partnership. Partnership has come a long way in recent years. More partnerships are now permanent and genuinely representative, and many participants have accepted the need for greater openness and accountability. In future, it is likely that further moves will be made to ensure the fuller participation of local communities in partnerships and to relate the operation of partnerships to the new mechanisms for government and governance that have now been established in Northern Ireland, Scotland and Wales, and which are emerging in the English regions. In addition, and most importantly, it is essential that partnerships continue to absorb the lessons of best practice from elsewhere. This 'on the job' learning is a matter of urgency if the quality and depth of partnership is to be further improved.

Postscript

As this book goes to press, a number of important new initiatives relevant to urban and regional regeneration are taking place. Chief of these are the establishment of the Regional Development Agencies, the work of the Urban Task Force and the Scottish 'experiment' in Community Planning. Each of these initiatives contains elements of the wider commitments announced by the present government in relation to social and community regeneration (especially in the New Deal for Communities), the increased emphasis placed on sustainable development (especially the environmental and social dimensions) and the revitalisation of local and regional democracy.

Particular aspects of these new areas of activity that are of importance include:

- the emphasis placed on the creation of more appropriate policies that reflect the difficulties and opportunities encountered in individual regions and localities, including the incorporation of the Single Regeneration Budget within the portfolios of the Regional Development Agencies and the greater emphasis now placed on area-based regeneration policy (Roberts and Lloyd, 1998);
- the realisation that there is a need to examine the problems of urban (and rural) areas in the round, rather than attempting to deal with individual issues in isolation; this is reflected in the appointment of the Urban Task Force, led by Lord Rogers of Riverside, that has a mission to identify the causes of urban decline and to support ways of establishing and realising a new approach to urban regeneration;
- the acceptance of the desirability of bringing together all relevant areas of activity and funding into a single territorial programme; this is what the Scottish Community Planning initiative is now introducing in the hope that such an approach will allow for the introduction of a process that will 'promote the well-being of communities' (COSLA and the Scottish Office, 1998, p. 5).

The above examples are illustrative of some of the new initiatives that have been introduced since the bulk of the work on this book was completed. They also illustrate the need for the continual renewal of thinking about the causes and consequences of urban and regional change, and for the adoption of an 'open' approach to the need to review and revise policy. However, whilst immediate 'facts' and procedures may change, many of the problems and opportunities encountered in the practice of regeneration are as enduring as the basic theories and methods that have been presented herein.

By the time that the second edition of this book is available, it is likely that urban and regional regeneration will have advanced further. Any ideas or suggestions for the content of the second edition will be welcomed by the editors.

Finale

Whilst this chapter may represent the end of this book, it also represents a contribution to the beginning of a new era for urban regeneration. Whilst much of what has been discussed in this book may be a matter of familiarity to many readers, it is hoped that the bringing together of theory and practice from many diverse sources will have added something to our collective knowledge and understanding. As to the future, perhaps the most appropriate final words for this text are those of Monika Wulf-Mathies in her introduction to *Europe's Cities*: 'our cities are a sea of potential which has not yet been tapped' (Commission of the European Communities, 1997).

Mobilising this potential in a sustainable and responsible manner is the future challenge for urban regeneration.

References

Beecham, J. (1993) Urban change: the local perspective, *Royal Society of Arts Journal*, Vol. 141, no. 5441, pp. 534–48.

Brotchie, J., Batty, M., Blakeley, E., Hall, P. and Newton, P. (1995) *Cities in Competition*, Longman, Melbourne.

Burton, P. (1997) Urban policy and the myth of progress, *Policy and Politics*, Vol. 25, no.4, pp. 421–36.

Carley, M. and Kirk, K. (1998) *Sustainable by 2020?* Policy Press, Bristol.

Chapman, M. (1998) *Effective Partnership Working*, Scottish Office, Edinburgh.

Cohen, M. (1996) The hypothesis of urban convergence, in M. Cohen, B. Ruble, J. Tulchin and A. Garland (eds.) *Urban Future*, Woodrow Wilson Center Press, Washington.

Commission of the European Communities (1997) *Europe's Cities: Community Measures in Urban Areas*, Office for Official Publications of the European Communities, Luxembourg.

Convention of Scottish Local Authorities and the Scottish Office (1998) *Report of the Community Planning Working Group*, HMSO, Edinburgh.

Geddes, M. and Martin, S. (1996) *Local Partnership for Economic and Social Regeneration*, Local Government Management Board, London.

Hall, P. (ed.) (1981) *The Inner City in Context*, Heinemann, London.

Jewson, N. and MacGregor, S. (eds.) (1997) *Transforming Cities*, Routledge, London.

Landry, C., Bianchini, F., Ebert, R., Gnad, F. and Kunzmann, K. (1996) *The Creative City in Britain and Germany*, Anglo-German Foundation, London.

Lawless, P. (1995) Recent urban policy literature: a review, *Planning Practice and Research*, Vol. 10, no. 3–4, pp. 413–18.

Loftman, P. and Nevin, P. (1995) Prestige projects and urban regeneration in the 1980s and 1990s: a review of benefits and limitations, *Planning Practice and Research*, Vol. 10, no. 3–4, pp. 299–318.

Oatley, N. (1995) Urban regeneration, *Planning Practice and Research*, Vol. 10, no. 3–4, pp. 261-70.

Organisation for Economic Co-operation and Development (OECD) (1990) *Environmental Policies for Cities in the 1990s*, OECD, Paris.

Pacione, M. (ed.) (1997) *Britain's Cities*, Routledge, London.

Parkinson, M. (1996) *Strategic Approaches to Area Regeneration*, European Institute for Urban Affairs, Liverpool.

Roberts, P. (1997a) Sustainable development strategies for regional planning and development in Europe: an ecological modernisation approach, *Regional Contact*, Vol. 11, no. 12, pp. 76–87.

Roberts, P. (1997b) Territoriality, sustainability and spatial competence, in M. Danson, M.G. Lloyd and S. Hill (eds.) *Regional Governance and Economic Development*, Pion, London.

Roberts, P. (1998) Regional Development Agencies: progress, prospects and future challenges. Paper presented at the Regional Science Association Annual Conference, York, September.

Roberts, P. and Lloyd, M.G. (1998) *Developing Regional Potential*, British Urban Regeneration Association, London.

Robson, B. (1995) Paying for it, *Journal of Planning and Environmental Law*, Special Issue on Politics and Planning.

Schnaars, S.P. (1987) How to develop and use scenarios, *Long Range Planning*, Vol. 20, no. 1, pp. 105–14.

Shaw, K. and Robinson, F. (1998) Learning from experience? *Town Planning Review*, Vol. 69, no. 1, pp. 49–63.

Thew, D., Holliday, J. and Roberts, P. (1982) *West Midlands Futures Study*, West Midlands County Council, Birmingham.

World Commission on Environment and Development (Brundtland Report) (1987) *Our Common Future*, Oxford University Press, Oxford.

Index